CW00538080

William Faden and Norfolk's 18th-Century Landscape

by

Tom Williamson and Andrew Macnair

WINDgather
PRESS

Windgather Press
is an imprint of
Oxbow Books, Oxford

© Tom Williamson and Andrew Macnair 2010
Reprinted 2013

All rights reserved. No part of this publication may be reproduced,
stored in a retrieval system, or transmitted in any form or by any means
(whether electronic, mechanical, photocopying or recording) or otherwise
without the written permission of both the publisher and the copyright holder.

ISBN 978-1-905119-34-9

A CIP record for this book is available from the British Library

This book is available direct from

Oxbow Books, Oxford, UK
(Phone: 01865-241249; Fax: 01865-794449)

and

The David Brown Book Company
PO Box 511, Oakville, CT 06779, USA
(Phone: 860-945-9329; Fax: 860-945-9468)

or from our website

www.oxbowbooks.com

Cover design by Sue White

Printed in Great Britain by
Berforts Information Press
Eynsham, Oxfordshire

Contents

List of Illustrations   v

Acknowledgements   vii

Abbreviations   viii

1   The County Maps   1

2   Faden's Map of Norfolk   49

3   The Map Redrawn   78

4   Commons, Greens and Heaths   100

5   Woods, Parks and Plantations   119

6   Fields, Farms and Fens   143

7   For Leisure and Edification   165

8   Traces of Antiquity   183

    Conclusion   200

Bibliography   201

Index   211

On DVD

Contents of the DVD and introductory notes

Appendix 1. List of county maps considered by the Royal Society
            of Arts and those given a prize

Appendix 2. List of county maps that have been analysed in any detail

Appendix 3. List of William Faden's county maps

Appendix 4. American maps by William Faden in the British Library
            and the Library of Congress, Washington

Appendix 5. List of errors found on Faden's Map of Norfolk

List of Illustrations

1   William Faden's map of Norfolk   2–3

2   Christopher Saxton's 1574 map of Norfolk   5

3   John Speed's map of Norfolk, 1611   7

4   Robert Morden's map of Norfolk, 1695   8

5   County maps published before 1750   10

6   County maps published between 1750 and 1775   10

7   County maps published between 1775 and 1800   10

8   County maps which won Royal Society prizes   10

9   The triangulation diagram on Joseph Hodskinson's 1783 *Suffolk*   11

10   Linked triangulation in central and north-west England   12

11   The publication of English and Welsh county maps between 1700 and 1840   13

12   A theodolite as used in 1780   26

13   Parliamentary enclosure of commons and wastes in England, 1750–1819   28

14   Parliamentary enclosure in Norfolk   29

15   Hills shown as 'woolly caterpillars' on Andrews' and Dury's 1773 map of Wiltshire   32

16   Thomas Malton's print of Charing Cross, 1795   53

17   An enlargement of Thomas Malton's print of Charing Cross, 1795   53

18   The Battle of White Plains, East New Jersey, 1776   55

19   Part of Richard Horwood's 1799 Map of London   59

20   An extract from Thomas Milne's London Land Utilisation Map of 1800   61

21   Jesse Ramsden's 'Great Theodolite'   63

22   Joseph Huddart's chart of the Norfolk Coast, 1790   65

23   Plan by John Watté of the proposed New Cut on the Eau Brink   66

24   The section of the Norfolk coast shown on the 1795 military survey   70

25   A section of the Norfolk coast in 1795   70

26   The same area of coast (as in Figure 24) as it appears on Faden's map   71

27   The Rotation of Faden's map shown by a true horizontal line   82

28   A Geo-rectified version of Faden's map confirms rotation   82

29   A Geo-rectified version of Faden's map compared to the original map   84

30   Changes in magnetic declination in England between 1250 and 2000   85

31   Eleven parishes in central-north Norfolk   87

32   Themelthorpe common, Foulsham   88

33   The principal triangulation points in England and Wales used in 1809   95

34   Dates of the Ordnance Survey drawings and revisions   96

35   Commons, heaths and soil types   104

36   The relationship between commons and parish boundaries   105

37   Commons around Attleborough and Great Ellingham   106

38   The relationship between types of common land and soil type   108

39 Thetford Warren Lodge   111

40 Rabbit warrens shown on Faden's map   112

41 Isolated churches   116

42 The distribution of different kinds of woodland and soil type   120

43 Sexton's Wood, Hedenham   121

44 Ovington and Saham commons, as shown on Faden's map   125

45 Fritton Common, as shown on Faden's map   127

46 Old oak pollards on Fritton Common   128

47 The distribution of landscape parks on Faden's map   130

48 Holkham Park, as depicted on Faden's map   131

49 Intwood Park, as depicted on Faden's map   132

50 A late eighteen-century map of Intwood park   133

51 The location of parks in terms of distance from Norwich   136

52 'Common fields' shown by Faden in the area to the west of Thetford   146

53 The distribution of 'common fields', unfenced roads and field   150
    boundaries shown on Faden's map

54 The distribution of open-fields enclosed by parliamentary   151
    act after 1795 in Norfolk

55 The distribution of 'barns' marked on Faden's map   152

56 The Great Barn in Holkham Park   153

57 'Marshland' – the siltlands comprising the northern part of the Norfolk 155
    Fens – as shown on Faden's map.

58 Halvergate Marshes, with scattered marsh farms and drainage windmills   157

59 The southern fens, as shown on Faden's map   160

60 The pattern of roads in Norfolk shown on Faden's map   166

61 The New Buckenham turnpike and the Ipswich turnpike near Norwich   168

62 Milestone beside the old A11 near Wymondham   169

63 The Aylsham Navigation, showing locks and mills   171

64 The distribution of pits, kilns, and brickworks shown   175
    on Faden's map

65 The location of wind and water mills, and their relationship   176
    to watercourses

66 The 'Old Oak' at Winfarthing, as shown on the tithe award map   185

67 Sites of gibbets shown by Faden, and their relationship   188
    with hundred boundaries

68 Hundred boundaries converging on the great Stock Heath   189
    in north Norfolk

69 Commons, hundred boundaries and parish boundaries in the   190
    Wymondham area

70 Variations in the size of medieval parishes in Norfolk   193

71 Stock Heath, Melton Constable park, and neighbouring commons   194
    and wood pastures

72 The distribution of wood pastures and ancient woods shown   197
    on Faden's map

73 The distribution of woodland in 1066, as recorded in Domesday Book   198

Acknowledgements

Particular thanks are due to Adam Longcroft, who originally suggested how valuable a digitised version of William Faden's map might be to students of Norfolk's landscape. Paul Laxton gave extensive and constructive advice on the County Map series, and Laurence Worms's knowledge of William Faden and the London map trade has proved invaluable. Raymond Frostick corrected important errors of fact concerning early Norfolk maps and Joe Nunn, previously of XYZ Digital Map Company of Edinburgh, gave advice on the geo-rectification of Faden's original map. Chris Barringer, who wrote a scholarly introduction to the facsimile of the map published in 1975, has also given much help and encouragement. Thanks also to Phoebe Fox-Bekerman, of the Royal Society of Arts archives; to Jon Gregory, Ian Hinton and Alan Ovendon at the University of East Anglia; and to David and Susan Yaxley; all of whom have provided important advice, help or information. Our gratitude also to Keith Bacon, Gerry Barnes, Sara Birtles, Patsy Dallas, Karen Morley, Andrew Rogerson, Nicola Whyte, Alison Yardy and other present and former research students at the University of East Anglia, whose researches have immensely informed our own work. Lastly, it will be clear from the bibliography and references that Brian Harley, who died in 1991, continues to have an overwhelming influence on the history of cartography and on this study in particular.

Abbreviations

BLO	Bodleian Library, Oxford
LRO	Lancashire Record Office
NRO	Norfolk Record Office
RSA	Royal Society of Arts
TNA: PRO	The National Archives: Public Record Office
YAS	Yorkshire Archaeological Society

CHAPTER I

The County Maps

....

The User of Maps ought to be at Pains to examine them …that so he may know
their Goodness and Defects. He should not take every Map that comes out, on
Trust, or conclude that the newest is still the best.

John Green, geographer, *The Construction of Maps and Globes*, 1717.[1]

Introduction

This short volume concerns the map of an English county – Norfolk – which
was published by a London map-maker called William Faden in 1797. The first
part discusses the context of the map's production and its place in cartographic
history, and describes the creation of a new, digital version of the map which
can be accessed online at www.fadensmapofnorfolk.co.uk (Figure 1). The second
part of the book details the various ways in which this electronic version can
be interrogated to throw important new light on the landscape of Norfolk, not
only in the later eighteenth century, but in much more remote periods of time.
The attached DVD contains over 150 maps which have been derived from the
digital version, and which illustrate many of the issues discussed in the text, as
well as being more generally useful to students of landscape history, historical
geography, and social and economic history.

The publication of the county maps of the British Isles in the eighteenth
century has been described as a 'cartographic revolution'.[2] Before examining in
more detail William Faden's map of Norfolk we need first to assess whether
this somewhat bold epithet is appropriate for the series of maps as a whole,
and for Faden's in particular. At its simplest level the history of county maps
can be divided into three periods, each overlapping with the next. The first
saw the small-scale county maps produced by Christopher Saxton, John Speed
and others which appeared in the sixteenth and early seventeenth centuries,
and which were repeatedly copied over the next two centuries. They were
followed, in a second period, after what appears now to be an inordinate delay,
by the one inch to the mile (occasionally two inches to the mile) county maps,
published between 1700 and 1820, and it is these that will mainly concern us
here. This was a period, essentially, of cartographic private enterprise. Well
over fifty county maps were published, some of which were of exceptional
quality when one considers the difficulties faced by the map-makers. The
cartographers' backgrounds encompassed the arts, mathematics and astronomy

FIGURE 1 *(overleaf)*.
William Faden's map of
Norfolk, 1797, digitally
redrawn.

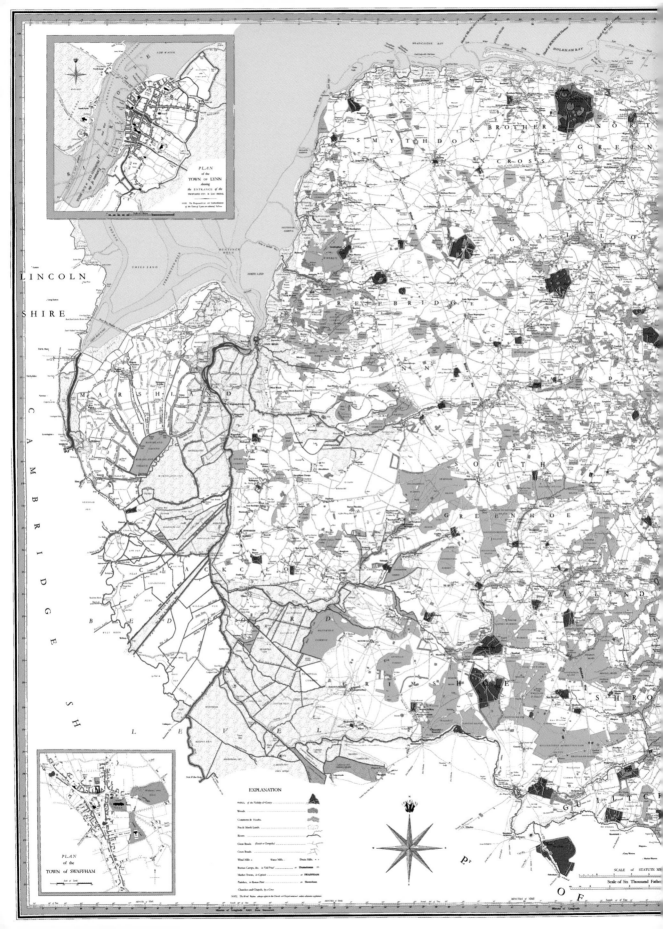

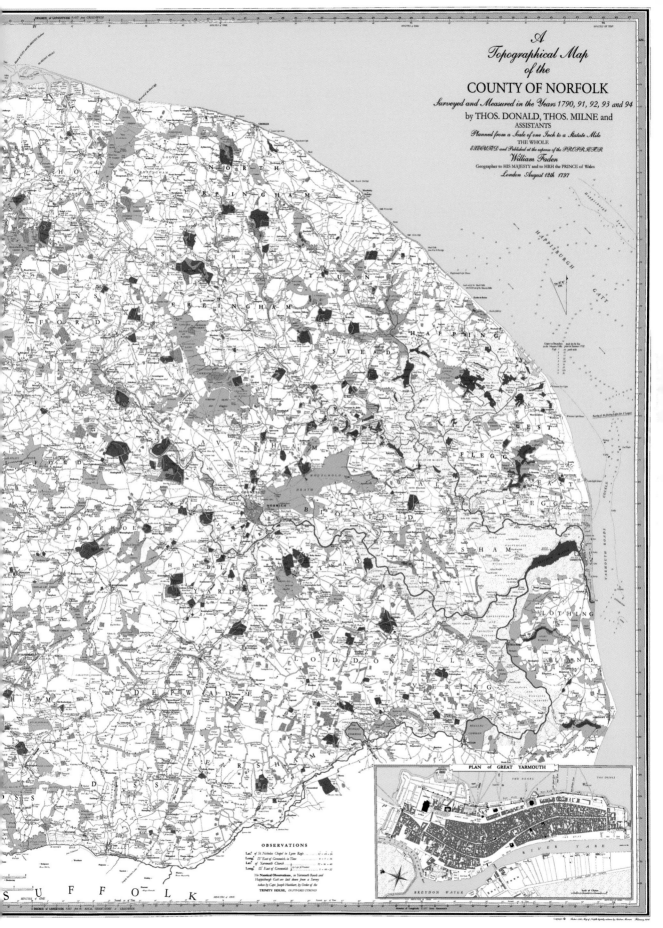

A
Topographical Map
of the
COUNTY OF NORFOLK
Surveyed and Measured in the Years 1790, 91, 92, 93 and 94
by THOS. DONALD, THOS. MILNE and
ASSISTANTS
Planned from a Scale of one Inch to a Statute Mile
THE WHOLE
EXECUTED and Published at the expense of the PROPRIETOR
William Faden
Geographer to HIS MAJESTY and to HRH the PRINCE of Wales
London August 12th 1797

PLAN of GREAT YARMOUTH

OBSERVATIONS

as well as mere surveying, and they worked and lived within families and learned societies steeped in the principles of the Enlightenment. This remarkable series was ended, in our third period, by the appearance of the Ordnance Survey. This organisation began to produce maps from the early nineteenth century onwards but only the first, that of Kent published in 1801, was based on a county format.

Saxton and Speed

Christopher Saxton (*c.*1542–1606), a Yorkshireman, embarked on his survey of the counties of England some time after 1570. He undertook systematic field work, travelling on horse-back, making compass sketches and using a plane table to plot rough maps.[3] Although he appears to have used a form of triangulation he depended essentially on survey by traverse – that is, taking lateral measurements from roads or tracks – while the Privy Council ordered on 10 July 1576 that Justices of the Peace should 'see him conducted unto any towre castle highe place or hill to view that countrey and that he may be accompanied with two or three honest men such as do best know the cuntrey … to safe conduct him to the next Market Towne'.[4] His initial survey of thirty-four counties took only five summers to finish and the very first map he completed was that of Norfolk in 1574 (Figure 2). This was, perhaps, not by chance. William Cunningham's 1558 map of Norwich is the earliest surviving printed map of any English town, and the appearance of both surveys may reflect the importance of the city, and the county, in the perceptions of contemporaries.[5] Saxton immediately sent copies to his supporter William Cecil, later Lord Burghley and Secretary of State, and to Thomas Seckford, a Suffolk lawyer and Master of the Court of Requests, who financed both the survey and the engraving. Queen Elizabeth knew of and supported Saxton's plans and in his rush to print the Norfolk map (engraved by Cornelis de Hooghe, a Fleming working in London) he initially forgot to include her coat of arms – an error which was rapidly rectified. By 1579 all the English counties had been surveyed: the maps were sold individually, as single sheets, for four pence each, and also gathered and bound together into an *Atlas of England and Wales*.[6] At the time it was an undertaking unparalleled in Europe (apart from Philip Apian's 1568 survey of Bavaria), although, surprisingly, Saxton omitted to put coordinates of latitude and longitude on his maps, the inclusion of which was by this time commonplace on the Continent. Brian Harley's view was that Saxton's survey was 'first of all an English expression of developments pioneered in Renaissance Europe'.[7]

John Speed (1552–1629) did not have the financial and political support available to Saxton and was thus obliged to make his series of county maps commercially viable. He did this in part by ensuring that they were made as visually attractive as possible, and while little or no fresh surveying was carried out, apart from the miniature town plans, he added appealing views of places within the county in question, as well as coats of arms, descriptive texts and

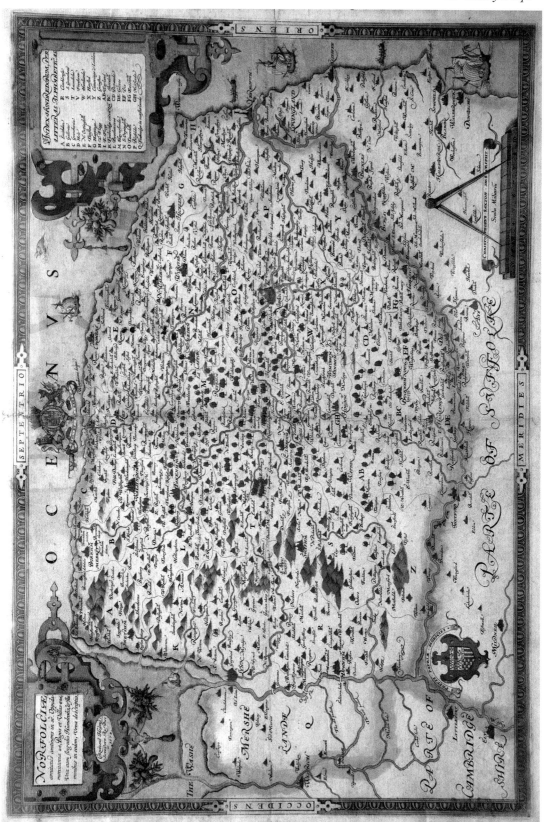

FIGURE 2. Christopher Saxton's 1574 map of Norfolk.

inset plans of the county town. He freely admitted that his maps were copies, stating: 'I have put my sickle in other mens corne'.[8] The first of the county map-makers to be based in London, he used the engraving skills of Jodocus Hondius in Amsterdam, though the maps were possibly returned to Speed himself for printing. His 1611 Norfolk map, part of *The Theatre of the Empire of Great Britaine*, was one of his first publications, and was presented as 'Norfolk a Countie florishing and populous described and devided' (Figure 3). Early in its publishing history the original copperplate for the map suffered a crack in the top left corner. Attempts were made in 1627 to repair this with riveting, which is apparent on subsequent maps. The prints continued to deteriorate and so a new plate was engraved in about 1665.[9]

For almost 100 years after Speed there were no significant developments in county cartography. Relatively few new surveys were undertaken and the Elizabethan and early Stuart maps were repeatedly reissued, copied and plagiarised; indeed, Saxton's map was still being printed as late as 1770, although this was perhaps consciously an antiquarian exercise. Richard Gough perceptively remarked in 1780 that 'as to the several sets of county maps professing to be drawn from the latest observations, they are almost invariably copies of those that preceded them'.[10] This lack of enterprise did not reflect any lack of interest in county maps; the map-publishing trade simply had no new material. As a consequence the Continental map trade, principally that in the Netherlands and France, left England and London in its wake. A few minor changes are apparent on those maps which were produced, however. John Seller's 1676 map of Hertfordshire, for example, used St Paul's Cathedral as the meridian or 'zero longitude'. Prior to this the Azores and Canaries were used as the meridian, as it was noted that there was no magnetic variation on compass readings when these islands were first being charted; it only later became obvious that magnetic variation changed over time, and that the negative readings were a chance observation. Robert Morden (1668–1703), whose business was based in Cornhill and whose county maps, including one of Norfolk, were published in Camden's *Britannia* in 1695, attempted to incorporate new antiquarian and topographical features, as well as the principal roads in the county based on Ogilby's road books[11] (Figure 4). But, again, no new survey was attempted.

The large-scale county maps

The eighteenth-century revolution in British cartography began in fits and starts.[12] Joel Gascoyne's map of Cornwall, published in 1699, was the first large-scale county map and, although it does not show conclusive evidence of being based on a trigonometrical survey, it almost certainly was. It is drawn at a scale of 0.84 inches to the mile; Gascoyne may well have intended to use a one-inch scale, but the need to fit the tapering peninsula onto the standard copperplate enforced the adoption of a smaller scale.[13] His map was followed

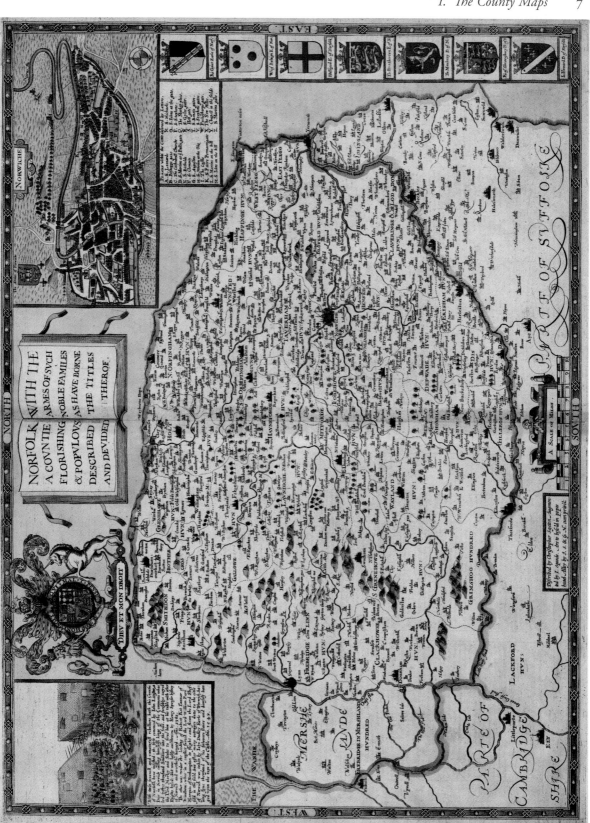

FIGURE 3. John Speed's map of Norfolk, 1611.

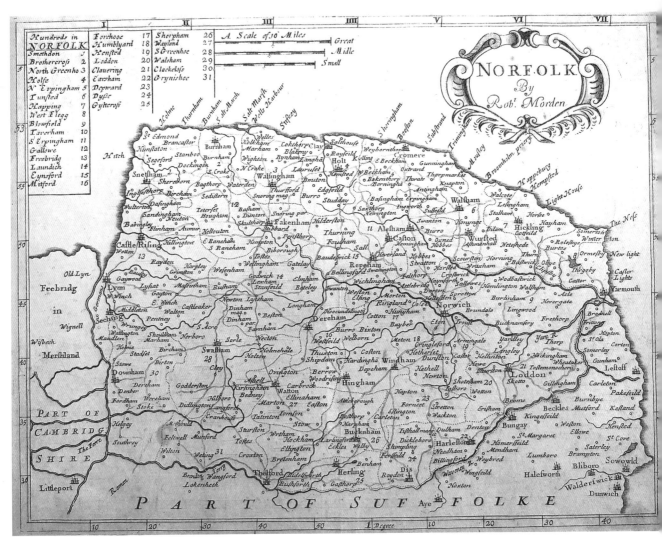

FIGURE 4. Robert Morden's map of Norfolk, 1695.

by Budgen's 1724 survey of Sussex, Senex's 1729 map of Surrey and Bowen's 1729 map of south Wales, all of which appeared almost a quarter of a century later.[14] Henry Beighton's 1728 map of Warwickshire was an exceptional piece of work, and the first to have been unquestionably based on a trigonometrical survey. It set the standard for future county maps.[15] The other maps produced in this period were not so good; Budgen's map of Sussex was described by one contemporary as 'deserves but the name of a map at most, and even as such is neither correct nor well executed', while Arthur Young thought John Kirby's 1736 map of Suffolk 'a miserable one'.[16]

In 1730 the estate surveyor James Corbridge produced the only large-scale map of Norfolk to appear prior to William Faden's, at a scale of three miles to two inches. Engraved in London by Emanuel Bowen, an apprentice-master and predecessor of Thomas Jefferys, it has recently been fully evaluated by

Raymond Frostick.[17] Like the abortive Mostyn Armstrong map of the 1770s (below, pp. 66–69), Corbridge's project was engulfed in controversy from the start. It went on sale for ten shillings but two prominent Norwich booksellers, William Chase and Thomas Goddard, seeing a financial opportunity, wrote openly to the *Norfolk Mercury* that the map had many errors and proclaiming that they would produce their own map (in fact no more than a pirated copy) and sell it for only eighteen pence. The Copyright Act of 1734 had yet to be passed. In further letters to both the *Norwich Mercury* and the *Norwich Gazette* a disgruntled assistant of Corbridge (with the unlikely name of Thermometer Elinett) hinted that a proper survey had not been undertaken.[18] The following week Corbridge, in an anonymous letter to the paper, vigorously denied the accusations and referred to Chase as a 'petty stationer'. The spat probably helped sales and it certainly did not stop the production of a further map by Goddard and Robert Goodman, an event which must have further confused a bewildered Norfolk populace. Corbridge's map itself, even though it was based on a new survey, is perhaps a disappointment; its topographical detail cannot be compared to Faden's map of sixty years later.

Indeed, it was only after 1750 that maps of a consistently high standard, and based on new field surveys made with improved instruments, trigonometrical methods and the latest mathematical and astronomical knowledge, began to appear. Figures 5–8 show the countrywide distribution of new surveys made before 1750, between 1750 and 1775, and in the last quarter of the century. By 1775 nearly half the country had been covered and by 1800 only Cambridgeshire[19] and the highlands of Scotland and Wales remained to be surveyed. In a frenetic fifteen-year period from 1765 to 1780 twenty-five English county maps were published, covering 65 per cent of the total area of the country. Survey by triangulation became the norm and eight of the county maps boasted of their surveyor's skills by including triangulation diagrams on the margins of the map (William Yates's maps of Lancashire, Staffordshire and Warwickshire, Peter Burdett's of Derbyshire and Cheshire, John Prior's of Leicestershire, Andrew Armstrong's of Durham and Joseph Hodskinson's of Suffolk (see Figure 9)). Some surveyors anticipated the Ordnance Survey by extending their triangulation into several adjacent counties (as shown on Figure 10). Yates in Staffordshire thus 'shared' prominent trigonometrical points, such as the Wrekin and Mow Cop, with Peter Burdett, working in the adjacent counties. Burdett himself, in his proposals for the Lancashire map, asserted that 'we shall contribute *our* Part to a more large and extensive one, a correct Map of the whole Kingdom – which we must not expect to see, till an actual Survey has thus been made of every County.'[20] In other instances the surveyor's methods were explained in a note on the map; Thomas Donald's map of Cumberland thus included the information that 'the Horizontal Distances are deduced by Trigonometrical Calculations from a base line measured on level ground.'

The period between 1815 and 1840 is unusual in that it saw the production of many privately financed new county maps, principally by the Greenwoods

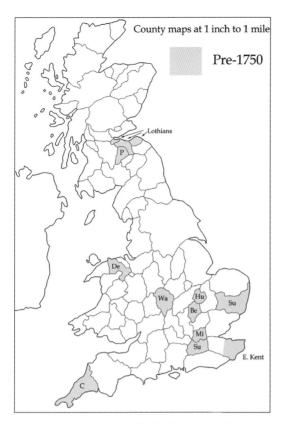

FIGURE 5. County maps published before 1750 (based on Hodgkiss in Harley 1965).

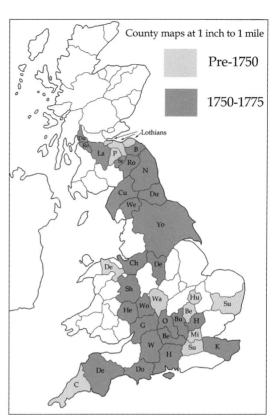

FIGURE 6. County maps published between 1750 and 1775 (based on Hodgkiss in Harley 1965).

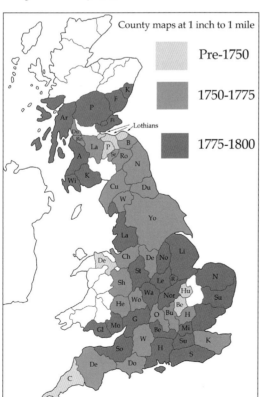

FIGURE 7. County maps published between 1775 and 1800 (based on Hodgkiss in Harley 1965).

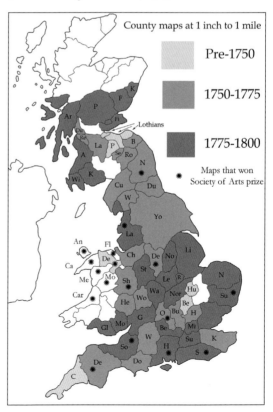

FIGURE 8. County maps which won Society of Arts prizes (based on Hodgkiss in Harley 1965).

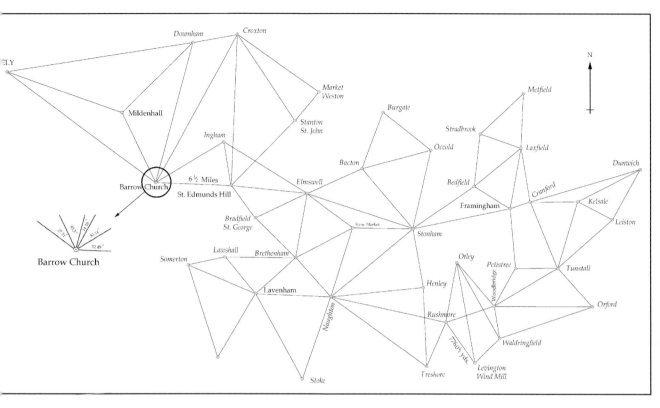

(Christopher and his younger brother John) and Andrew Bryant. They were surveying at the same time as the new Board of Ordnance, from whom trigonometrical data was made freely available. Both the Greenwoods and Bryant had ambitions to map the whole of England and were more professional in their approach than earlier surveyors had been; their maps in many ways equalled in quality those produced by the Ordnance Survey. The Greenwoods were almost successful in their aim, completing all but six counties, one of which was Norfolk: Bryant managed only thirteen, although he did produce a map of Norfolk, on which he was able to include parish boundaries. Figure 11 shows how the production of county maps changed over time: the high volume between 1815 and 1835 reflects this late flowering of the Greenwoods and Bryant's cartography. If this group of maps is ignored (many are of counties which had already been surveyed at a large scale and were thus not really 'new') we can see that Faden's map of Norfolk is a comparative late-comer in the sequence. Indeed, Norfolk was the last English county to have a newly surveyed large scale map, if we ignore Cambridgeshire.

The Board of Ordnance

It would be a mistake to think that there was a sudden improvement in the quality of surveying and map production at the end of the eighteenth century with the arrival on the scene of the Board of Ordnance. The private county map-makers

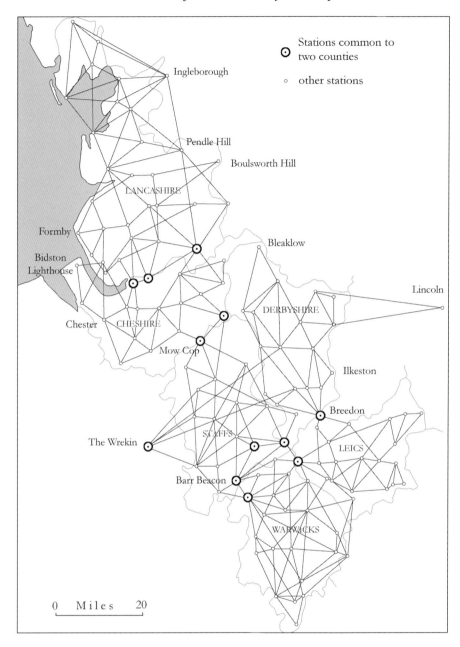

FIGURE 10. Linked triangulation in central and north-west England, from the county maps of P. Burdett, J. Prior and W. Yates (based on Hodgkiss in Harley 1965).

had mastered nearly all the skills needed to create maps of a very high standard, even if at times they lacked coordination and consistency. Indeed, between 1795 and 1820 the private and public map-makers exchanged much trigonometrical and other cartographic information, as well as members of staff. The Trigonometrical Survey of the Board of Ordnance (only from 1854 termed the Ordnance Survey) was set up in 1791, although triangulation and topographical work had begun in 1784 in the Plymouth area.[21] As its name suggests, the Board of Ordnance was a

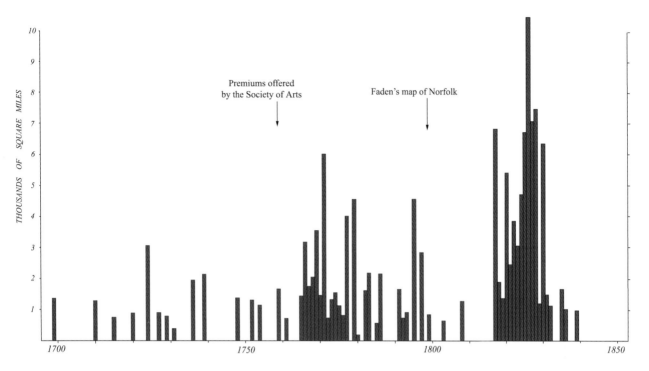

FIGURE 11. The publication of English and Welsh county maps between 1700 and 1840. Each column represents the total area covered by all the maps published in that year. The county areas are taken from the 1851 census (after Laxton 1976).

supplier of munitions connected to the Royal Arsenal at the Tower of London, and map production at the Drawing Room was a subsidiary part of its role. Its survey of the southern counties started in 1792 and trigonometrical data was published between 1799 and 1811, and made available to the various private map-makers still working on county surveys.[22] The first map produced by the Board, of Kent, was surveyed from 1797 and published in 1801 by William Faden. The survey was supervised by William Yeakell, who had previously worked on a map of Sussex published in 1795 (and likewise by Faden). It was no accident that Kent was the first county to be mapped by the Board: England anticipated hostilities with France at the time and 'the county, from its proximity to the coast, may in process of time, become the seat of military operations'.[23]

Trigonometrical data from the Board of Ordnance had been employed earlier, however, for the 1795 map of Sussex. This was based on an earlier and partial survey of the county which had been made by Yeakell and Gardner in 1783 and was completed after a lapse of twelve years by Thomas Gream.[24] Joseph Lindley and William Crosley had also used data from the Hounslow Heath base-line measurement, made in 1784, for their 1792 map of Surrey, and this should perhaps be considered the first use of 'official' data in a county map. The work of the Board of Ordnance will be considered in greater detail below, as the activities of this body provide us with an important context for Faden's Norfolk map.

An interesting pattern of state involvement in cartography becomes apparent when we compare the three periods of county mapping. Saxton's map was undoubtedly a government survey. It was promoted by the Crown as an act of policy, on the advice of the Queen's ministers, and used for national administration.[25] Saxton was given a 'placart', or pass, which was signed by ten Privy Councillors and enabled him to gain access to any part of the realm. The later Board of Ordnance maps were also, of course, 'official' surveys. Sandwiched between these enterprises were the eighteenth-century county maps, all of which were essentially products of private enterprise, and it is these which are our chief concern here.

A 'cartographic revolution'?

The first question we need to address is why there was such a deluge of new county maps in the second half of the eighteenth century: why, that is, was 70 per cent of England's area surveyed between 1765 and 1783? A little simplistically, perhaps, we might attribute this upsurge of activity to a number of interconnected factors. First and foremost was the growing complexity of the national economy – indeed, the emergence for the first time of a truly integrated national economy, and the demands this placed on transport infrastructure, manifested in such things as the development of turnpike trusts from 1750 and the canal frenzy of the 1790s. Between 1730 and 1780 publications like the *Gentleman's Magazine* and the *London Magazine* were full of plans for new waterways and harbours. But improved transport systems, for people and goods, by road, sea and navigable rivers and canals, all demanded better maps. Second were more ephemeral influences – questions of national identity, pride and defence, ranging from the military threat from across the channel to simple rivalry with France, which had already embarked on a scheme for nation-wide mapping in 1759. The lack of a similar initiative at home was difficult to justify given that there was no lack of skilled surveyors, well used to working on estate surveys, and that London was already a major centre for map engravers, printers and publishers. Above all, perhaps, this new cartography was in a sense, a late expression of the scientific principles of the Enlightenment, which not only replaced religious belief and superstition with observation and experiment, but also saw accurate knowledge of the world as the necessary precondition for changing it. With the development of improved surveying instruments, these principles could now be used to determine the shape of the earth and accurately portray its surface. In numerous and complex ways, the new county maps both portrayed and were themselves a product of the process of modernisation occurring in later eighteenth-century Britain.[26]

National aspirations were, however, refracted through the lens of more local affinity and identity: for the county was still the major unit of administration, politics and elite social interaction. And it was at the county level, too, that money was available to finance new maps, by the raising of subscriptions from the county elite; no financial help could be expected from central government. Expertise in surveying was also available locally; many of the private county maps were produced by individuals from within the immediate community.

Yet it was at this level that national aspirations were constantly being thwarted. In September 1755 the Rev. William Borlase (1695–1772), an antiquarian from Ludgvan near Penzance, wrote to his friend Henry Baker (1698–1774) – naturalist, antiquarian and Fellow of the Royal Society:

> I would submit to you as a Friend, whether the State of British Geography be not very low, and at present wholly destitute of any public Encouragement. Our maps of England and its counties are extremely defective … and the Headlands of all our Shores are at this Time disputed … if among your premiums for Drawings some reward were offered for the best Plan, Measurement and actual Survey of City or District, it might move the Attention of the Public towards Geography, and in time, perhaps, incline the Administration to take this matter into their Hands (as I am informed it does in some foreign Countries) and employ proper Persons every year, from Actual Surveys, to make accurate Maps of Districts, till the whole island is regularly surveyed … and attempts what would be so much for the Honour as well as Commerce of this Island'.[27]

It was the Society for Arts (more fully the Society for the Encouragement of Arts, Manufactures and Commerce, and from 1847 the Royal Society of Arts) that focused national and local concerns about the inadequacies of county maps.[28] Henry Baker was himself a founder member of the Society, which met for the first time in 1754 at Rawthmell's coffee house in Henrietta Street, Covent Garden. From its inception the Society had a keen practical interest in transport, and encouraged reliable surveys of roads and coasts. Unlike the Royal Society it was not exclusively a learned society – it was more of a technical association, serving the growing world of trade, commerce and the industrial arts.[29]

The Society's members were well aware that Louis XV had commissioned Cassini de Thury to map the whole of France on a trigonometrical framework, and the Society debated for several years how best to encourage a similar enterprise in Britain before deciding 'on the utility of bestowing a premium for the best map of any District or County in the Kingdom made from an actual measurement and survey'.[30] The first advert appeared in 1759, offering a prize of £100 for an 'accurate, actual survey for which the Society would procure satisfactory proof of its merits'. It noted the great value of such a project 'for the Improvement of Highways, making rivers Navigable and providing other means for the Ease and Advancement of the National Commerce'. By this time it was accepted that a full national survey was not feasible. Instead, two or three counties should be surveyed at a time, a proposal later changed to 'any county'.[31] Over the next seven years the advert for the prize varied slightly but included at various times the following:[32]

1. A stipulation that the maps should be based on a trigonometrical survey.
2. Specifications of which instruments were to be used for the survey: 'the horizontal distances of all places in the map to be taken with the theodolite with a plain [in later surveying texts spelt *plane*] table and the roads to be measured with a perambulator and noted down in figures.' 'Taking of Angles by Circumferentor or such like uncertain instruments is disallowed but new invented instruments of known use and certainty are allowed.'

3. An instruction that the maps should be drawn at a scale of one inch to the mile and should include measurements of road distances.
4. A time scale for completion – at most, two years were allowed to complete and deliver the survey (a stipulation which was later relaxed).
5. The offer of an additional gratuity 'for an exact and accurate level and section of rivers that are capable of being made navigable'.
6. The instruction that 'Sea coasts of all maritime counties … be correctly laid down, together with their latitude and longitudes'.
7. A statement that the maps would be 'most carefully assessed, certificates of accuracy would be required of the country gentry and examination would be made of the trigonometrical data, and field books'.

Within weeks, the first advert published drew a response from Benjamin Donn (1729–1798), a teacher of mathematics and natural philosophy from Bideford in Devon, who wrote to the Society stating that he would start on his survey of that county 'immediate on receipt of the necessary instruments … which [are] expected from London very soon'.[33] His survey took five and a half years to complete, longer than he expected but understandable as he had an occupation other than that of map-maker, and he duly received his prize of £100 in 1765. The first submission to the Society was in fact made by Isaac Taylor of Ross on Wye for his Dorset map, but no prize was given and the Society minutes do not state why (although Richard Gough, a firm critic of maps, commented that 'the map is very faulty in the place names').[34] The second prize, again of £100, went to Peter Burdett's 1767 map of Derbyshire. Burdett was summoned to the Society of Arts because his map was not drawn exactly at the requested scale of one inch to the mile; he responded that the map had warped during production ('in Printing off from the Plate on Paper there is a small contraction')[35] and his prize was confirmed. He, too, was an amateur surveyor; it is noteworthy that of the first nine unsuccessful applicants to the Society three were professional cartographers and three were full-time estate surveyors.

Between 1759 and 1809 thirteen prizes or premiums were given for new county surveys, eleven from England and two from Wales. Only twenty-three maps were entered for recognition by the Society of Arts out of the more than fifty published: a full list is given in Appendix 1. William Faden was involved, as publisher, in three of the prize-winning maps: that of Suffolk surveyed by Joseph Hodskinson in 1783 (himself a member of the Society and the first professional surveyor to receive an award); that of Hampshire surveyed by Thomas Milne in 1787; and that of Sussex, surveyed by Thomas Yeakell, William Gardner and Thomas Gream in 1795. Faden did not receive any recognition for his Norfolk survey, a matter to which we shall return.

The Society's patronage was of crucial importance in fostering the new cartography.[36] This was not so much because of the actual monetary value of the prizes given to the map-makers: the total value of all the prizes combined was only £460 (plus 4 gold medals, 3 silver medals and 'one great silver pallet with suitable inscription' given to Charles Masters, assistant surveyor of Somerset), a sum which would not have been sufficient to cover the costs of producing any single map. It was more that, while the Society failed in its early attempt to encourage a national survey, it did succeed in bringing some standardisation to the private county maps. Its insistence on checking triangulations and field books, and on seeking independent referees for the surveys, established the parameters for the subsequent production of the Ordnance Survey maps.

The county maps: an overview

A comprehensive review of the fifty or so county maps which made up this 'cartographic revolution' has yet to be written, and it is thus difficult to assess the place of Faden's map of Norfolk within this group. Indeed, it is only since the 1960s that the significance of Britain's eighteenth-century large-scale mapping has been recognised, mainly through the writings of J. B. Harley.[37] The production of commentaries and facsimiles of the various county maps is still in its early stages, and has mainly been carried out under the aegis of local historical and archaeological societies. No more than a dozen (Appendix 2) have been studied in any detail – less than a fifth of those published – and the quality of some of the facsimiles can at best be described as adequate: in the future, modern digital photography and the archiving of maps on compact discs and other digital media should result in easier access to good-quality images. All this said, it is clear that the influence of the Society of Arts in setting an overall standard did not prevent the production of highly individualistic maps. The idiosyncrasies result from a variety of factors: the topography of the county in question, the unique series of events involved in the production of each particular survey, the history of publication and republication and, on occasion, the personal histories and characters of the map-makers concerned. None of the maps and their associated documents can answer all the questions we would like to ask about their production and *raison d'etre*. Indeed, it is a truism that, apart from the maps themselves, remarkably little evidence has survived concerning their production. What follows is thus a synopsis, derived from many disparate sources.

In this brief summary of the British county maps those produced for Scotland and, in particular, Ireland have received less attention than they perhaps deserve. One of the county maps of Wales does warrant a mention, however, particularly because of the contrast it presents with the experience of William Faden and his Norfolk publication. John Evans published his *Map of the Six Counties of North Wales* in 1795. It was highly thought of, and seven years later he received a premium of forty-five guineas from the Society of Arts. Evans was a gentleman and amateur cartographer who lived in the small village of Llanymynech, six miles south of Oswestry. His map was engraved not by a professional engraver but by the parish clerk of the village, Robert Baugh, and it was printed locally in Oswestry.[38] As we will see, the contrast with Faden's more sophisticated and extensive establishment in London could hardly be greater. Robert Baugh was so enthused by the success of his Welsh map that he went on to engrave and publish one of Shropshire; this also received a prize from the Society of Arts, in 1808, in the form of fifteen guineas and a silver medal. The letters sent to the Society in support of the Shropshire map included one from Thomas Telford, then the Surveyor of Works in Shropshire, and another from Robert Waring Darwin, the father of Charles.[39]

It is important not to see the British county maps in isolation. They were part of a pan-European phenomenon and London was only one of several important centres for map-making, the others being Amsterdam, Nuremberg and, above all, Paris.[40] Pedley has written that 'Paris and London were the epicentres of Enlightenment science; their respective academies (the Royal Society and the Académie Royale) supported and promulgated geodetic and geographic studies'.[41] William Faden, as we shall see, eventually acquired the title of Geographer[42] to the King, but the French equivalent, the *Géographes du Roi*, had much greater social and academic status and even received government funding.[43] England was particularly fortunate in that the district around Soho in London became a home for refugee Huguenot copperplate engravers, who brought with them new techniques from across the Channel.[44] French stylistic influence can be recognised in several of the county maps, probably because several surveyors, including John (Jean) Rocque (Surrey and Berkshire), Peter (Pierre) André (Essex) and Andrew (André) Dury (Hertfordshire), had a Huguenot background.[45]

The purchasers

It was a map-maker's clientele, those who bought his finished product, who ultimately decided on the success or failure of a map, and the cartographer's attempts to keep these people happy were a major influence on the kinds of topographical feature which were included. The county maps were not cheap, but unlike their French counterparts they could be afforded by all who were reasonably well-off.[46] Nevertheless, at a time when the Board of Ordnance surveyors were paid between 4s and 7s 6d per day, engravers and journeymen surveyors 6s a day and a Captain of the Foot in the army 10–15s a day, the three guineas necessary to purchase the Norfolk map (admittedly one of the more expensive) would have made most people pause.[47] We can only guess at the motives of the purchasers. Were the maps bought with ideas of land improvement in mind, especially enclosure? Did county administrators find them of value (they were certainly employed when decisions were made concerning county boundaries for the first national census in 1801)? The Board of Agriculture evidently found them of some use, and those of counties lying along the south and east coasts were of service to the military authorities as the threat from France became real. As postage rates were charged by the mile, those showing the location of mile posts were clearly of value to the Post Master General. But, perhaps more importantly, the maps would have been of general interest to the educated public, to the kinds of people who also purchased the various county histories that were being written in some numbers at the time. Some of the maps were doubtless valued as decorative items, works of art, a 'source of visual pleasure':[48] even before the first county had been surveyed maps are described as 'the most commodious ornament for every man's House'.[49] Above all, members of polite society liked to see their status confirmed, with their names proudly emblazoned beside their houses and estates, large and small.

We are fortunate in that for some of the maps we know from the subscriber's list the names, titles and even addresses of purchasers. A list naming the 264 subscribers for John Prior's 1777 map of Leicestershire has thus survived in Richard Gough's collection in the Bodleian Library in Oxford. It includes, as we might expect, members of the nobility, gentry and clergymen, but also embraces surgeons and physicians, Masters and Fellows of Oxbridge colleges, attorneys, schoolmasters, surveyors and many private individuals, middle-class people without known status or title.[50] Most were residents of the county or lived within fifty miles of its boundary, but a significant number were from more distant parts of the country. Although a full analysis of subscribers to county maps has yet to be undertaken, there is evidence that over the course of the eighteenth century the proportion of subscribing nobility gradually fell while the number of 'professionals' (clergy, medical men, lawyers and military officers) rose. The number of 'non-gentry', including craftsmen, instrument makers, industrialists and, in particular, booksellers, also rose.[51] Map collectors account for some of the county map sales. A review of Sotheby's sales between 1790 and 1840 shows us that Michael Lort, Regius Professor of Greek at Cambridge, sold twenty county maps in 1791, while in 1814 the Duke of Norfolk bought 'a large parcel of English county surveys'.[52]

The surveyors

The two men actually responsible for surveying Faden's map of Norfolk, Thomas Donald and Thomas Milne, were both surveyors by profession, and employed full-time on the project, circumstances which were by no means the norm in the production of the county maps. Peter Burdett, who surveyed the maps of Derbyshire (1767) and Cheshire (1777), was an artist of great talent who has been described by some as, if not the inventor of aquatint engraving, then certainly the first to introduce the technique to England.[53] Andrew Armstrong (Durham (1768), Northumberland (1769) and several Scottish counties) was originally a soldier by profession,[54] while Richard Davis (Oxfordshire, 1797) was a local farmer and agricultural writer and William Yates (Lancashire, 1786 and Staffordshire, 1775) a customs officer. The latter even advertised the sale of his county maps from his workplace, the Liverpool Customs House.[55] It was a time when men moved easily from one field of study to another.[56]

Benjamin Donn (1729–98) published his map of Devon in 1765 while employed as a teacher of mathematics and natural philosophy in Bideford, then a busy seaport sending ships to Holland, France and Virginia in the Americas.[57] He was a regular contributor to the *Mathematical Repository*, producing reports on the timing of eclipses in Bideford and London. As well as writing his acclaimed *Mathematical Essays* he contributed articles to the *Gentleman's Magazine* on calculating the moon's parallax and on changes in the refractive index in the eye following cataract surgery.[58] His five years' involvement with the map of Devon was followed by a move to Bristol, where he set up a Mathematical Academy. In his later life he was appointed Master of the Mechanics to His Majesty.

John Prior (1729–1803), who produced the 1777 map of Leicestershire, was a clergyman in Ashby-de-la-Zouch and for forty years schoolmaster in the grammar school there.[59] He had particular interests in mathematics, surveying and musical theory and was a competent violinist. Francis, Earl of Huntingdon, to whom the county map was dedicated, supposedly said of him that he was 'a poor teacher – but you should hear him fiddle'. Joseph Whyman, a former pupil, did much of the surveying for the map, having been taught by Peter Burdett, who was responsible for the 1767 map of Derbyshire. Prior and Burdett knew each other and the trigonometrical surveys of the two counties were to some degree coordinated. The Society of Arts awarded him twenty guineas and a silver medal for his map, having initially decided on a gold medal but changing its mind; it could be a fickle organisation. After publishing his map Prior retreated into obscurity and continued to work as a schoolmaster right up to his death. His obituary in the *Gentleman's Magazine* noted his 'perfect freedom from ambition, love of music and relish for the calm delights of literature and domestic society'.[60]

Joseph Lindley worked for only two years of his life on his map of Surrey, published in 1792. An extremely able mathematician, he had previously been a London banker (working for Prescott, Grotta and company) but later entered the Time Department at Greenwich Observatory where he assisted Nevil Maskelyne, the Astronomer Royal, in his work on the determination of longitude.[61] Lindley originally hailed from Yorkshire; for some reason many of the surveyors came from Scotland or the north of England, and many had a nonconformist, often Quaker, background. There were virtually no Oxford or Cambridge graduates involved in the profession of surveying. Several were trained at the various Dissenting Acadamies set up in the 1750s and 1760s by men whose religious convictions barred them from attending the universities. These, together with endowed schools, tended to concentrate on the teaching of mathematics as well as on technical and vocational subjects.[62] The Society of Arts seems almost to have encouraged the involvement of non-professionals, whose varied talents and direct involvement in the affairs of their immediate locality undoubtedly gave an extra dimension to their maps. This is not to say that professional surveyors were never involved, however. Isaac Taylor of Ross on Wye was an estate surveyor before he became involved in no less than five county maps, while Thomas Yeakell and Thomas Gardner were both professional surveyors who later worked at a senior level for the Board of Ordnance. Henry Beighton, who brought out his Warwickshire map in 1728 (one of the earliest of the large-scale county maps), was likewise a surveyor by profession, although he had many other talents, writing in the *Philosophical Transactions* of the Royal Society on his theories of perspective drawing (the Society made him a Fellow for his design of a steam pumping engine, 'Beighton's fire engine', which was used to remove water from collieries).[63]

The surveyor was not always a welcome figure. To the farmer and tenant he was associated with rents, taxes and tithes, and on many occasions individual members of the profession had stones thrown at them or were refused horses. The Rev. Alexander Bryce, while surveying the Caithness coast, was told that the locals 'were not anxious that navigation should be made safer lest it should deprive them of the spoils of the numerous wrecks along the coast'.[64] Nor was it any easier across the Channel. The map-maker Cassini de Thury wrote 'the course of my travels in Brittany nearly cost me my life'; his men were stoned and their equipment stolen, they were accused of being spies, his father was shot at and locals refused to furnish horses or guides.[65] He was particularly upset that fellow countrymen abused him as 'in foreign lands my name alone draws respect'.[66] None of this was new. Two hundred years earlier, in 1587, Christopher Saxton was driven from Broughton Hall in Yorkshire by the sitting tenant, Henry Tempest, his two sons and a Spaniel dog.[67]

The publishers

We have seen that some of the surveyors were able to publish county maps on their own account. In general, however, the business of publication was monopolised by the London map-makers. They had the great advantages of also being, in many cases, engravers, printers, colourers and retailers, and of being located in the capital, where they were in a position to purchase at auction houses the copperplates of earlier maps, which could then be re-engraved and reprinted. Many had extensive business interests. Thomas Jefferys, whose firm was eventually acquired by William Faden (see p. 50), was involved in the publication of ten county maps and a number of detailed town plans prior to his death in 1771, but these constituted only one part of his business career. He published a wide range of other material and his cartographic interests were truly international.

Despite the remarkable number of county maps produced in this period only a relatively small number of publishers, surveyors and engravers were involved, mostly based in London, working in competition but at the same time sharing in particular enterprises. This was partly to lessen financial risk and partly to allow them to pool particular skills: engraving and publishing (as with Thomas Jefferys and William Faden); surveying and publishing (Joseph Hodskinson) or surveying and engraving (John Rocque and Joel Gascoyne).

It was common for surveyors to move from publisher to publisher, as well as moving frequently around the country and into Scotland. By Faden's time map-makers had begun to migrate away from the city area, around St Paul's and the Royal Exchange, to the more fashionable West End.[68] Andrew Dury, who produced maps of Hertfordshire, Kent and Wiltshire, worked in Duke's Court just off St Martin's Lane, very close to William Faden's shop: so too did the Scotsman Patrick Begbie, the engraver of the Masters' map of Somerset. John Andrews, a surveyor closely associated with Dury, was based at nearby 29 Long Acre.[69] Joseph Hodskinson, who worked with both Jefferys and Faden, was located in Arundel Street, the Strand, where John Cary the map-maker also had his first shop (at No. 188). John Rocque moved to Old Round Court, also in the Strand, after a fire destroyed much of the stock in his previous shop near Charing Cross.

The geodetic survey

In the early years of the eighteenth century Norfolk was indirectly associated with an event that radically changed the arts of navigation, surveying and cartography as a whole. On 22 October 1707 Sir Cloudesley Shovell, originally from Cockthorpe on the north Norfolk coast, was returning to England in his flagship the *Association* using Grenville Collins's 1693 chart of the Scilly Isles in his *Great Britain's Coasting Pilot*.[70] Unfortunately this incorrectly placed, by nine minutes of latitude and nine nautical miles, a group of rocks, an error compounded by the fact that Shovell's navigators misgauged their longitude position. The *Association*, the *Eagle* and the *Romney* foundered on the rocks and more than 2,000 sailors were drowned. It was one of the British fleet's greatest disasters and led to the passing of the Longitude Act in 1714: this established the Longitude Board which, over the ensuing 110 years, stimulated, by the offer of substantial financial prizes, improvements in nautical and terrestrial surveying.[71]

Advances in instrumentation and mathematical theory from the sixteenth to the eighteenth centuries, given further encouragement by the Longitude Act, finally allowed areas as large as counties to be accurately portrayed on the surface of the earth. This involved *geodetic survey* – the establishment of an accurate mathematical template – followed by *topographical survey* – the positioning of objects such as churches, roads or rivers within it.

John Ainslie wrote:

> The Surveying of a large district or county is an operation so extensive and complicated, as to require the utmost exertion of a surveyor's abilities in every branch of his department; for in the exercise of it, he will find various difficulties to encounter.[72]

Any survey of a large area involved first setting out divisions, or *graticules*, of latitude and longitude, based on astronomical measurement. When these had been determined, trigonometrical methods were used to fix individual points within each graticule. We do not have a comprehensive description of the methods used in the survey of any of the county maps but we do know the various text books which were available to, and used by, their surveyors. One was *Geometrical and Graphical Essays*, published by George Adams in 1791, two chapters of which were written by Thomas Milne, one of the surveyors involved in Faden's map of Norfolk.[73] Later in date but equally important were John Ainslie's *The Gentleman and Farmer's Pocket Companion* (1802) and his more detailed *Comprehensive Treatise on Land Surveying* (1812). John Ainslie was unarguably Scotland's outstanding surveyor of the late eighteenth and early nineteenth centuries and was well acquainted with Thomas Donald, the other surveyor involved in the production of Faden's Norfolk map, having worked with him on several county surveys. Ainslie was apprenticed to Thomas Jefferys in 1762 and often published in association with William Faden. He produced his books late in life, after working extensively on county maps, and we can be confident that his book was a genuine reflection of late eighteenth-century practice. He was well aware that the problems facing the county surveyor were, as noted above, greater than those encountered in the making of simple estate surveys.

Before starting on a county survey, two major decisions had to be made. Firstly, the surveyor had 'to chose a proper piece of ground whereon to measure a base'; and, secondly, he had to decide on 'stations', or prominent intervisible points, that would form the basis of his triangulation of the county. The base line had to be level and its length measured as accurately as possible using a Gunter's chain, which was 22 yards long and divided into 100 links (it was named after Edmund Gunter, Professor of Astronomy at Gresham's College in 1620).[74] The Dee estuary was chosen as the base line for Burdett's map of Cheshire, while William Yates described how, for his Lancashire survey, 'two Base Lines, one of six and the other of ten miles were carefully measured on the Sea Beach'[75] (the exact position is not noted but probably included the sandy shores to the north of Liverpool). For his map of Warwickshire, Henry Beighton utilised straight sections, as long as eight miles, of the Roman roads the Foss Way and Watling Street. An alternative was to set out several base lines of a shorter length, an approach adopted by Andrew Armstrong for his 1768 map of Durham: in this case, most were less than four miles in length.

Once the base lines had been set out the first angular measurement was made, using a theodolite, from the end of a line to one of the chosen stations. In a field book the 'principal bearings' were noted, as well as 'a circle of bearings all round'.[76] The instrument was then moved to the next point on the line, the observations repeated, and 'back bearings' made. The exercise continued until all the principal stations in the county had been included. A minority of county maps include a diagram showing the principal stations and the relevant bearings. Joseph Whyman's map of Leicestershire (surveyed in 1775–6) has nineteen stations and forty-two rays, while Yate's of Staffordshire (1769–75) has twenty stations and fifty-one rays. The average length of sight between the principal stations in Burdett's map of Cheshire was 9 miles, very similar to the 8.3 miles in the case of Leicestershire and the 11.9 miles in Staffordshire.[77] Much greater distances were sometimes used, however. For his Lancashire map Yates sighted Coniston Old Man from Ingleborough – a distance of 33 miles – although even this was less than Burdett's 48-mile sighting of the Wrekin. Telescopic sights would have been essential for such observations.

The county map surveyors would have been aware of the need to set the county within the correct latitude and longitude in order to ensure that the maps were correctly located in a national system of coordinates. Determination of latitude was relatively easy – it could be established from the altitude of the sun at noon using a quadrant and tables giving the sun's declination on that day. The calculation of longitude was a greater problem; in essence it had to be reckoned by comparing local time with time at the meridian (usually Greenwich Observatory). Since the earth rotates at a steady rate of 360° per day, or 15° per hour, there is a direct relationship between time and longitude. A time piece or chronometer was set at the meridian and compared with local time when, using a quadrant, the sun reached its highest point in the sky (local noon). John Harrison's series of five chronometers, built between 1736 and 1773, greatly improved longitude measurement in maritime surveys but they were also of value in making terrestrial maps.[78] By the late eighteenth century it had become technically possible to measure longitude to within a few seconds of arc, but few county map-makers felt that this degree of accuracy was necessary. The borders of some county maps were graduated in degrees of latitude and longitude (for the latter often using the crude measure of 69½ statute miles to one degree), but it is likely that this was an addition made at the engraving stage.

The meridian, the 'zero line of longitude' based at Greenwich Observatory, was used by Thomas Jefferys in his series of county maps and, of course, a little later by the Ordnance Survey. Earlier maps had used a local meridian: Shrewsbury in the case of Rocque's map of Shropshire, Winchester on Isaac Taylor's of Hampshire, and Salisbury Cathedral on Andrews' of Wiltshire. For some of the county maps, such as Burdett's survey of Cheshire, it is not clear what meridian was chosen: possibly St Paul's Cathedral or the Cornhill in London. There is no unequivocal evidence that any of the county map-makers considered the problem of projection – that is, the potential errors which arise when the curved surface of the earth is represented on a flat piece of paper. John Seller and Robert Morden, both cartographers, wrote on mathematics (and Seller also on navigation)[79] and it is unlikely that they were unconcerned about the matter.[80] However, the scale of any distortion was fairly minor when dealing with the kind of area covered by an English county, although it does mean that maps of Norfolk and Suffolk, surveyed independently of each other, would not exactly match at their contiguous edges.[81] Perhaps we should not expect the county surveyors to have been overly concerned with details of latitude, longitude, meridian and projection, however, as they were of only marginal interest to those wishing to purchase the maps. Surveyors with a mathematical background such as Benjamin Donn (Devon) and Joseph Lindley (Surrey)[82] took this part of their survey particularly seriously, and others tried to satisfy the 'curious in Geography'.[83] But in general this was an aspect of the county surveys not always as accurately covered as topographical details.

The topographical survey

Once the main triangles had been plotted the surveyor had to fill in the details of the landscape using other methods. John Ainslie succinctly described the best way of going about this:

> First, I would advise the surveyor to make himself well acquainted with the district or county he is about to survey. This may be effected by riding or walking over it in various directions with a person well-informed of every particular part. By this means the names of the several towns, villages, seats, hills and other remarkable objects, may be obtained, which will prove of the most essential service in the progress of the survey.[84]

It is likely that many surveyors performed a secondary triangulation – 'Inferior parts were drawn in a like Manner' – using prominent features, such as churches and windmills, within the principal triangles;[85] Ainslie describes elsewhere in his manual how 'those triangles are again filled up by smaller ones'.[86] In his map of Surrey Joseph Lindley lists the 85 principal stations of his angular survey, together with 430 other objects 'the situation of which was settled by the foregoing observations' (though he comments that the smoke of London 'totally obscured the objects in Surrey').[87] Some surveyors over-emphasised the geodetic side of their maps to the detriment of the subsequent topographical drafting and, as a result, their maps, while technically as accurate as possible for the time, were less pleasing to the eye. Among the most sparsely detailed maps are those by Burdett, Prior, Armstrong and Donn; yet all four were considered worthy of prizes by the Society of Arts.

The details within the secondary triangles might be plotted by using a plane table, a portable surveying instrument consisting of a drawing board and ruler mounted on a tripod. Observations were plotted directly onto the map by the surveyor in the field. But this procedure, although favoured by Jefferys in the 1760s, was very time-consuming and it is likely that the lesser triangles were usually filled in by road traverses, with the distances recorded by a perambulator or measuring wheel. Henry Beighton (Warwickshire) designed and built a perambulator thirteen feet in circumference, far larger than the standard instrument employed. He even used a calculation to compensate for 'irregularities in the surface of the road'.[88] Benjamin Donn states that for his Devon map he took an average of ten theodolite measurements along each mile of roadway. Further measurements for single buildings, cross roads and so on would probably have been made with the circumferentor, a small theodolite, even though its use was not favoured by the Society of Arts. Sketches of hills and other features were entered in field books which had later to be interpreted by the draughtsman and engraver. For some county maps there is ample evidence of fairly liberal interpolation of topographical detail between points that had been fixed with a theodolite.

The Society of Arts wished the county surveys to be completed and delivered within two years, but this time limit was often exceeded and it was not an absolute bar to winning a Society award. Day and Masters's map of Somerset, published in 1782, took seven years to complete,[89] Donn's of Devonshire took five and a half years, while Yates's of Lancashire was published a full eleven years after the initial proposals. Lindley's map of Surrey was surveyed in nine months, a remarkable achievement, but he directly copied parts of Rocque's earlier map, inadvertently admitting to his plagiarism when he recorded payments to the engraver to undertake the copying.[90] He also used a number of existing estate surveys, including one of Richmond Gardens which was twenty years out of date.

Surveying instruments

The very high quality of the county maps produced in the second half of the eighteenth century was due in great measure to the steady improvements made in the design and construction of surveying instruments during the seventeenth and early eighteenth centuries, as well as to major advances in the theoretical analysis of mathematical error in instrumental readings. Of particular importance was the availability of theodolites capable of taking bearings over long distances correct to one minute of arc (one sixtieth of a degree) (Figure 12). John Ainslie advised surveyors to use 'a good theodolite ... at least 6 or 7 inches in diameter'.[91] As already noted, the Society of Arts was opposed to the use of the circumferentor, a surveying compass provided with sights for taking horizontal angles, which was reputed to be subject to errors of up to two degrees.[92] The two vital improvements in instrumentation which occurred in the course of the eighteenth century were the appearance of the achromatic lens (a lens designed to limit the effects of chromatic and spherical aberration), said to have been invented in 1733 by the English barrister Chester Moor Hall, and first patented by John Dollond in 1785; and the development of techniques which allowed instruments to be manufactured with graduated mechanical scales.[93] The latter were perfected by Jesse Ramsden (a colleague of William Faden in the Smeatonian Society) who in 1773 invented the circular dividing engine. Prior to this, the division and inscription of scales on mathematical instruments were carried out by hand with an accuracy no better than 3 seconds of arc. Ramsden's divider made it possible to inscribe lines accurate to one second, and mechanisation of the process meant that instruments could be produced more rapidly and were cheaper and lighter. In 1785 Ramsden commenced construction

FIGURE 12. A theodolite as used in 1780.

of his Great Theodolite, an instrument which featured a horizontal circle three feet in diameter and which was used in the Greenwich–Paris triangulation of the late 1780s. With a weight of 200lbs, this instrument was of course far too large to be employed in the surveys for the county maps, but Ramsden's more general influence in improving the accuracy of surveying cannot be doubted. The base line for the Greenwich–Paris exercise was on Hounslow Heath (now partly under Heathrow airport) and for this General Roy used rods of glass rather than (as was more usual) deal, because their expansion could be accurately gauged using a pyrometer, which was again designed by Ramsden.[94] Other important instrument-makers in London at this time included John Bird and the brothers John (another Smeatonian) and Edward Troughton.[95] By the end of the eighteenth century, as a consequence of such technical developments, England led the way in the design and manufacture of surveying instruments – even if the French excelled in national mapping, as well as in geodetic and topographical surveying. But these instruments were extremely expensive: in 1764 a theodolite with a vertical arc and telescope divided to every minute cost £30, while George Adams estimated in the 1790s that it would cost £125 to equip a surveyor with all the instruments he needed.[96]

Topographical detail on the county maps

To modern eyes the most striking differences between the landscape depicted on the county maps and that shown on the modern Ordnance Survey maps are the massive increase in the extent of built-up areas and the disappearance of vast tracts of common, heath and other 'wasteland'. In some northern counties the former was accompanied by (and was indeed a consequence of) large-scale industrialisation in the course of the nineteenth century. Even counties that have remained largely rural have seen significant growth, often through more localised industrial and commercial expansion. Leicestershire had a population of about 115,000 in 1777, the year in which John Prior's map was published, a figure which had increased sevenfold by 2000.[97] In contrast, the population of Norfolk, 273,000 at the time of the 1801 census (shortly after William Faden's map was surveyed), had increased only threefold by 2000.[98] The disappearance of 'wastes' was largely a consequence of the parliamentary enclosures which occurred rather unevenly over the whole country, mainly in the period between 1750 and 1830. As Turner and others have shown, these took place in two great waves: one peaking around 1780, which principally affected the main areas of open fields in the clay vales of the Midland counties; and the second during the Napoleonic wars, which affected remaining areas of open arable on the light soils of eastern England, but which was mainly concerned with the removal of heaths, moors and other areas of common land.[99] Although enclosure by parliamentary act was the final stage in a long process of eradicating intermixed arable land and land used and managed in common – replacing both with hedged or walled parcels, individually owned or occupied – it had a profound and dramatic effect on both

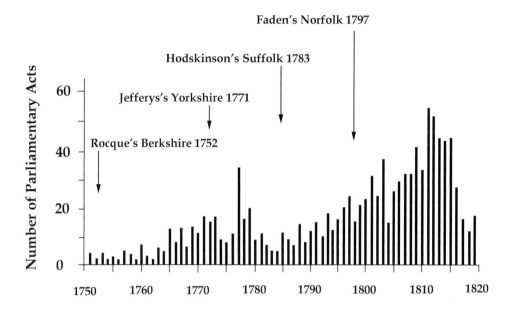

FIGURE 13. The chronology of parliamentary enclosure of commons and wastes in England, 1750–1819 (after Turner 1980).

landscape and society.[100] The extent to which the various county maps reflect this change depends on three main variables: the date at which the map in question was surveyed; the extent to which the county had already been enclosed; and the amount of open common land, as opposed to open arable, which was removed by parliamentary acts, for surveyors generally mapped commons more accurately, and more completely, than open fields.[101] Figure 13 shows the chronological distribution of enclosure acts in England and Wales involving 'commons and wastes' during the late eighteenth and early nineteenth centuries; superimposed are the dates of some of the county maps.

We have already noted that Faden's map of Norfolk came quite late in the sequence of county maps. But as most of the parliamentary enclosures in the county occurred during the second 'peak', it essentially shows the situation at an early stage in the process, and the majority of the common land which had existed in the county in the later Middle Ages probably remained at the time the map was surveyed. For this reason, if for no other, it is of immense importance in the study of landscape history.

The point is reinforced by Figure 14, which shows the chronology of parliamentary enclosure, of open fields and commons, in Norfolk alone. The next large-scale county map of Norfolk, published by Bryant in 1826, shows the landscape after enclosure had been largely completed. In contrast, Leicestershire was a county of extensive open fields on heavy soils and with relatively little common land. It was largely affected in the first wave of parliamentary enclosure and so, although its county map was published in 1777, twenty years before Faden's Norfolk, it shows a landscape where 60 per cent of the commons and heaths had already disappeared. As a result it is inherently less interesting for those concerned with the history of the landscape.

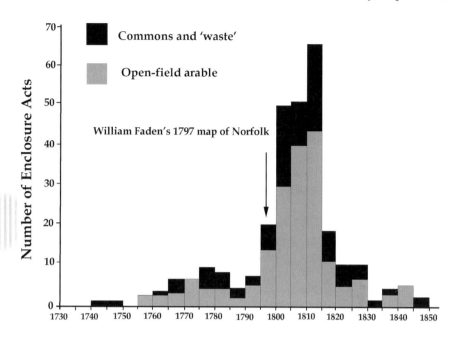

FIGURE 14. The chronology of parliamentary enclosure in Norfolk (after Turner 1980).

The purchasers of the county maps were less concerned with geodetic than with topographical detail, and the majority of adverse comments made about particular examples were concerned with such matters as incorrect spelling, poor legibility, lack of colour, missing features or the inclusion of outdated information, particularly in relation to roads and transport. Nor should we forget price; customers of John Strachey's 1736 map of Somerset were unhappy at the price of 7s. 6d.: 'too much for two sheets of paper'.[102] Map size, the quality of engraving and the general aesthetic 'feel' of the map were probably the main concerns of most potential purchasers. In marked contrast to the later Ordnance Survey maps, the county surveys had no uniform way of showing topographical detail: indeed, it is the resulting diversity of presentation that gives the county maps their unique appeal to modern users and purchasers. The Society of Arts, as we have seen, did insist on certain elements being recorded, but this still left the surveyors, draftsmen and engravers enormous leeway in what they included, and how. Some surveyors, such as John Rocque, had a background in estate surveying and brought elements of their experience in this field to their county maps; others, like Isaac Taylor, had a particular interest in antiquities, and his Dorset map accordingly included plans of various Iron Age hillforts. The earlier county maps quite commonly record old fields of battle and encampments, in a way that echoes those produced much earlier by Speed. Later these historical features were rather frowned on, perhaps because they did not reflect the serious scientific endeavour of the Enlightenment. Faden's map of Norfolk has none of the antiquarian or picturesque embellishments often found on earlier maps; there are no cartouches, heraldic shields or scenes of country life. Instead we see a map created primarily for practical people, in a world of industry and agricultural improvement.[103]

The makers of the county maps seem to have held widely differing opinions as to what topographical features should, and should not, be included, and for this reason we should be wary of comparing them directly with later Ordnance Survey maps. For the same reason we cannot expect county maps, when analysed collectively, to provide us with an accurate picture of the distribution of any kind of landscape feature. Yeakell and Gardner's unfinished map of Sussex (1778–1783) was surveyed at two inches to the mile, and was 'by far the most detailed and carefully surveyed of all the county maps but it did not include parish boundaries and made little or no attempt at industrial features'[104] (only the southern half was completed, in four sheets, partly because only 250 of the required 400 subscriptions were forthcoming, and Yeakell and Gardner were transferred to the Board of Ordnance).[105] John Rocque's maps of Berkshire, Middlesex and Surrey were unusual in that they were published at a scale of two inches to the mile, and clearly reflect the surveyor's previous interest in parks and estates as a *dessinateur des jardins*. His 1761 map of Berkshire, like Andrew's and Dury's survey of Hertfordshire and Davis's map of Oxfordshire, also provides some indication of current land use. This practice has been criticised by later commentators, on the basis that individual fields and their use could hardly be represented with any accuracy at this scale. Paul Laxton has, however, shown that Rocque, although technically wrong in his drafting of some completely fictitious field boundaries, did faithfully portray the overall balance and distribution of arable and grassland in Berkshire.[106]

The boundaries of the ancient administrative units called 'hundreds' were shown on virtually all the county maps, in spite of the fact that their practical importance was, by this time, fading rapidly. Parish boundaries were shown on only a few. Joel Gascoyne's 1699 map of Cornwall shows all 201 parishes in that county, as does Thomas Martyn's 1748 survey of the same county; John Rocque's 1754 map of Middlesex and 1761 map of Berkshire likewise show them. Given the speed at which some surveys were carried out it is remarkable that the recording of parish boundaries was ever even considered, particularly as in some areas parishes were intricately intermixed. It was only in the nineteenth century that they were routinely included on commercial maps (such as Bryant's 1826 survey of Norfolk); by the end of the century they had been comprehensively recorded on the one-inch Ordnance Survey maps. The depiction of designed landscapes – the parks and gardens of the nobility and gentry – must be viewed with some caution. In Yorkshire, for example, the county map of 1771 anticipates planned changes in several places, such as at Burton Constable, where the alterations made by Lancelot Brown are clearly recorded although they were not in fact executed until 1774.[107]

The depiction of roads was generally taken seriously by the surveyors. Donn recorded that in measuring the roads of Devon his perambulator travelled over 6,000 miles, and Burdett's map of Cheshire has 2,700 miles of road plotted. Major and minor roads were sometimes differentiated by engraving the former with a thicker line on one side. Turnpike roads were frequently shown, as were turnpike gates, tolls, milestones and road distances, although, once again, the maps sometimes anticipated reality. Enclosed and open roads are distinguished, as usually are roads enclosed on one side and open on the other. Faden's Norfolk map appears to distinguish between the two, but the map key is not as explicit on this matter as are those on most other county maps.[108]

Relief was usually, although not invariably, shown by means of hachures. The results and accuracy are variable; many of the fells in Donald's Cumberland are missing and substantial peaks, such as Great Gable and Glaramara, are not named.[109] Tourism, Coleridge and Wordsworth had not reached the county in 1774 but within fifteen years this had all changed and Donald's resurvey in his *Environs of Keswick* provided far more detail for this new group of customers. Occasionally the heights of particular hills are given, but not usually accurately. That of Pendle Hill, for example, is given as 1,568 yards in Jefferys' Yorkshire, and Misty Law in Armstrong's Ayrshire at 1,430 yards, against the true heights of 1,827 and 1,562 feet respectively. The depiction of relief had been a challenge for map-makers from the start. Saxton used humps or 'mole-hills' to show hills and these were still employed, in only a slightly more sophisticated way, on some of the county maps, including Andrew Armstrong's survey of Ayrshire, where they are admittedly used to good effect. The 1773 map of Wiltshire is typical of the 'woolly caterpillar' style of hill representation, which can easily exaggerate the real variation in relative relief (Figure 15).[110] The use of hachures to depict slope was first developed in the mid seventeenth century by the Swiss map-maker Hans Gyger, but was initially kept as a military secret.[111] Only in the eighteenth century were they widely employed, with the general rule 'the darker the steeper, the darker the higher'. The limits imposed by copper engraving to some extent influenced the use of hachures, but the county maps can boast some particularly fine examples of the technique. One of the problems with most published facsimiles of the maps is that the hachures tend to become exaggerated in the copying process, with consequent loss of adjacent detail. The use of contour lines to show elevation was standard on the Continent from the 1790s but was not introduced into England (by the Ordnance Survey) until the late 1830s and was first used in the case of Norfolk only in 1887.[112] The first use of contour lines in Scotland, in contrast, came in 1777,[113] when the mathematician Charles Hutton and Rev. Nevil Maskelyne, Astronomer Royal, were attempting to calculate the mean density of the earth by detecting the attraction of the Scottish mountain Schiehallion, in Perth and Kinross, on a plumb-bob: Hutton connected together 'by a faint line all the points of the same relative altitude' in order to measure the dimensions of the mountain and thus calculate the earth's mass.[114]

The county maps were published when many districts were already experiencing extensive industrial development, yet before the growth of the great conurbations of the nineteenth century. In general, the mining and metallurgical industries are less well recorded on the English county maps than on their Continental counterparts: they have been closely studied by industrial archaeologists and found

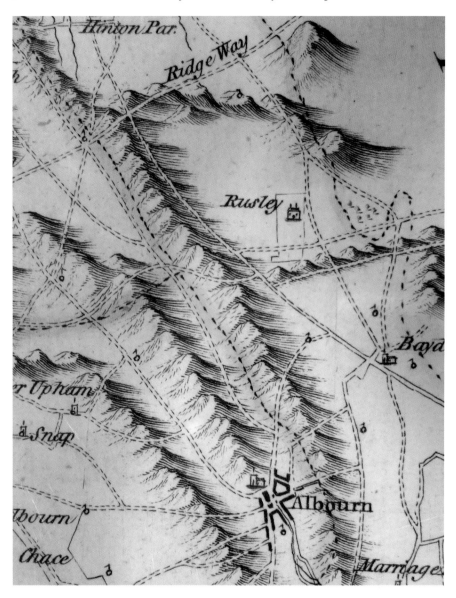

FIGURE 15. Hills shown as 'woolly caterpillars' in the area around Aldbourne on Andrews' and Dury's 1773 map of Wiltshire.

wanting in many respects.[115] In general, windmills seem to be well represented and some maps, including Faden's Norfolk, seem to show virtually all that existed. Watermills, in contrast, are undoubtedly under-recorded, and their depiction within a county is often patchy.[116] Boulton and Watt's steam-driven mills (both men were Smeatonians, like Faden) were in use by 1778 but they are not recorded as such on any of the maps apart from Armstrong's Ayrshire. The maps provide a notoriously poor record of the coal and iron industries, even those of Lancashire and Yorkshire, where (for example) the coal pits near Barnsley and the furnaces at Knottingley are omitted.[117] The county of Durham boasted the most active coal field in the world when Andrew Armstrong surveyed it in 1768, but studies using contemporary maps and documents show that his map provides a very partial representation, in part because the same symbol was used to represent an individual pit, a colliery with several pits or a general area of mining activity. There is some evidence that in Yorkshire coal pits belonging to particular owners, for example the Marquis of Rockingham, are shown more comprehensively than others, perhaps simply because these men or their agents provided the necessary information.[118] Despite evident omissions the Lancashire map won a Society of Arts award, and we should perhaps remember, both in this case and others, that coal pits were simply too numerous, and in some cases too small, for all to have been recorded at this kind of scale. Other forms of industry also received somewhat cursory treatment. In his map of Cheshire Peter Burdett thus seriously under-recorded the salt industry (74,000 tons of salt were exported from the area each year, and salt pans would have been visually very obvious)[119] as well as omitting the silk mill in Congleton, which employed 600 people. Similar observations on the incomplete recording of industrial and mining sites could be made about Burdett's map of Derbyshire.[120] Yates's maps are particularly unreliable in their depiction of industrial sites; his 1775 Staffordshire map does show Erturia, Joseph Wedgewood's model pottery factory built between 1769 and 1771, but provides no indication that Burslem boasted a major pottery industry. The same map shows a canal that was planned, but never built: so too does Jeffery's Yorkshire map, which includes sections of the Leeds to Liverpool canal which were never constructed. William Faden, in what was for him an unusual lapse, likewise records on his revision of Burdett's Cheshire map a branch of the Ellesmere canal that was in fact never completed.

Some map-makers, however, do appear to have been more concerned with industrial enterprises. Prior's 1777 map of Leicestershire thus notes 'fire engines' in a number of places – steam-driven pumping engines, 'engines for raising water by fire'. On Beighton's Warwickshire map slitting mills, boring mills and blade mills are noted and differentiated (the latter an early sign of the developing sword and cutlery industry based on Birmingham). Indeed, the county maps remind us of some industries, locally important at the time, which are now only a dim memory. Cheshire, Somerset, Staffordshire and Ayrshire all had 'burning coal seams' and 'coal wagon ways', the precursors of the railways. Perhaps needless to say, the county maps are silent on domestic industry: nothing on the Leicestershire map, for example, hints that the town of Leicester was home to over a thousand knitting frames.

Drafting, engraving and printing

It is easy to overlook the importance of the draftsman in the production of the county maps; but he acted almost as a cartographic editor. He would have been presented by the surveyor with the results of all the field work: plane-table sketches, angular measurements and calculations, details of traverses in field notebooks; and from all of this he had to produce his initial drawings. Having decided on how many sheets were needed to cover the whole county, he would then have plotted the principal triangulation points and added a rectangular graticule. He then had to decide on a scale which would fit this fixed framework. Faden, for his Norfolk map, chose a scale of 'statute' miles, otherwise known as 'measured' or English miles – the unit that was to become the standard mile of 1,760 yards (this was shorter than the 'geographical' 'British' or 'Scots' mile – generally of 1,984 yards, but in fact variable in length – which was employed on some county maps).[121] A scale of one inch to the mile had become the standard after John Ogilby's 1675 *Britannia* road maps, and was widely used in military circles. The county maps often note that one degree of longitude was equivalent to 69½ statute miles and 60 geographical miles (as in John Prior's 1777 map of Leicestershire). The draftsman had to ensure that the map did not become too crowded with detail, and it is quite likely that some features noted by the surveyor had to be omitted. It was also easy for transcription errors to creep in at this stage of a map's production.

Engravers had an equally important role, that of producing something that people would actually wish to buy. These individuals were usually based in London, although Peter Burdett had his Cheshire map engraved by Thomas Billinge in Liverpool, as did Yates his map of Lancashire. As the quality of the map would depend so much on the skills of the engraver it was common practice, in the larger map offices, to employ different engravers to do specific tasks: one to draw hachures, for example, one the waves of the sea, one to provide the elaborate cartouches and compasses, one to do the lettering – although different people would sometimes provide Roman and Italic. Outline details – of coasts, roads and the perimeter of features like woods – were usually engraved first. Names were added next, then ornamental features such as trees within woods, and finally the hachures.[122] Intaglio engraving, which involved cutting into the copper plate with a burin, produced only lines; etching, in which a design was burned into a plate with caustic chemicals, produced a more textured print which had the disadvantage of wearing badly with repeated printing.[123] For his maps of the American Revolutionary War, William Faden introduced the use of the roulette, a small wheel with spikes on the circumference, to show stippling. This may have been used to show commons and heaths on the Norfolk map.[124] The copperplates used were of two types, red and yellow, and a continuum in between depending on how much zinc was added as a hardner. The yellow plates, with a greater zinc content, were harder and engraving with a burin in consequence more physically demanding: but they had the advantage that more copies could be printed from each plate, an important consideration given the price of copper, especially during the period of the French Revolution when it rose dramatically because of its use as a defensive measure on the bottom of ships.[125] The copperplates were very heavy; each of the six plates used for the Norfolk map would have weighed about 30lbs (14kg).[126]

The skills of the engravers varied. Patrick Begbie, who was responsible for Day and Masters's 1782 map of Somerset, was considered to be 'competent but not top-flight'.[127] Burdett's engraving of Cheshire appears hurried and his failure to depict landscape parks or provide the names of landowners beside significant properties probably reflects the fact that he moved to Saxony before the map was finished.[128] Certainly, his map can hardly stand comparison with those produced by Yates,[129] Jefferys, Rocque, Hodskinson, Andrews and Dury. Isaac Taylor's maps of Hampshire and Dorset likewise have a rather 'scruffy' engraving finish ('clumsy, Unequal Manner of the Engraving' according to John Hutchins, the Dorset historian, in 1780[130]) but this is compensated for to some extent by his interest in antiquities, especially Iron Age forts. Isaac Taylor used etching extensively so his merit should really be judged on good early impressions. The engraving plates should ideally have been checked by both surveyor and draughtsman; Mostyn Armstrong was unable to travel down to London from Scotland to inspect the plates of his 1775 Peebleshire survey, and admitted this in his *Companion* to the map.[131]

When the copperplate had been completed, and was ready for printing, it was heated by a brazier, inked and then wiped over. This left ink only in the cuts, and a final impression was taken by passing a dampened sheet of paper through a rolling-press under great pressure (lithography, a more rapid method of printing, was only invented in 1796 and did not become commercially available for twenty years after the appearance of Faden's Norfolk map).[132] Proofs would have been pulled from the press at various stages in order to check on the progress of the work and it is likely that these were circulated as specimens within the county.[133] An average of 100 impressions could be taken in a single day, sufficient for about sixteen complete copies of Faden's map of Norfolk.[134] The number of map-prints that could be taken from a single plate depended on many factors, including the softness of the copper, the quality of the paper and the skill of the printer. The burin lines on the plate coped well with repeated pressing: the more delicate etching showed signs of wear much earlier. Comparison of two versions of the same county map thus allows an estimate to be made of how early in the run each was printed. Between several hundred and three thousand prints could be taken from a copper plate before quality deteriorated beyond an acceptable standard. One important factor in this was the way the plate was cleaned of excess ink after inking. If a rag soaked in urine (*au chiffon*) was used the process was quicker, but the acid eventually damaged the plate. If the palm of the hand was employed (*á la main*), the job took longer but the plate lasted for a greater length of time.[135] The specialised ink used in map printing consisted of a pigment called German Black suspended in nut oil. This was made in Frankfurt from 'the dregs of burnt wine' and its use was continued by the Ordnance Survey.[136] Press-work was a printing craft in its own right, and the Board of Ordnance initially farmed out their printing to Cox, Son and Barnett, copperplate printers of No. 6 Breams Buildings, Chancery Lane.[137] We do not know if Faden printed maps on his own premises or whether the work was put out to others.

The character of the paper on which the county maps were printed had an important influence on how well they were presented – and on how well they have survived.[138] Papermaking was undergoing major changes at the time, as paper with 'chain' or 'laid' lines was replaced by woven paper, which had fewer blemishes. By the 1780s the manufacture of 'wove' paper was spreading quickly to papermills all over England.[139] Lack of finances meant that Burdett used poor-quality paper – not watermarked, and lacking even 'chain' or 'laid' lines – for his map of Cheshire, which consequently became warped, so that the scale was no longer true.[140] In contrast, Donn's map of Devon carries a Whatman watermark, indicating that a good-quality paper was used. William Yates offered to supply his Staffordshire map, at extra cost, on 'superfine paper, pasted on canvas': surviving copies have a fleur-de-lis watermark.

When they left the printing press the maps were damp, and drying could take as long as six weeks; if they were used before this time had elapsed the print could smear or the ink become impressed onto another map, if stored in a roll.[141] Warping and shrinkage after the drying process were a major problem; Yates noted that his Staffordshire map, expected to be 66 by 43 inches (1676 by 1092mm), was only 58 by 40 inches (1484 by 1028mm) after drying.[142] To compensate for shrinkage the engraver often made his original engraving larger by a fraction of one sixtieth.[143] Finally the county maps were issued either in loose sheets, or bound into large folio volumes, ideal for a gentleman's library. A tiny number were printed on vellum: more usefully, many were dissected, folded on cloth and sold in a small slip case.

Once the maps were complete, they were first delivered to those on the subscription list and then, after a delay of a few months, they were put on sale, at a greater charge than the subscription price, to the general public. Over subsequent years one or more plates were updated, by partial flattening (by pounding the copper from the back)[144] and re-engraving, and then republished. The revised map sheet was then sold as single sheets or as a set, sometimes perhaps with older unchanged adjacent sheets; the resulting collection would thus contain maps of several different dates.

Maps and copyright

Several of the county maps, although not Faden's Norfolk, either bear themselves, or have published in separate advertisements, the words 'Published 1st Jan. 1777 as the Act Directs', a reference to the Copyright Act then in force. The first Act of 1709, passed during the reign of Queen Anne, was essentially for the benefit of the book trade and gave control of the work to the bookseller alone, and not to the author.[145] More relevant to map-makers was the 1734/5 Engraving Copyright Act (widely referred to as 'Hogarth's Act' after William Hogarth, who lobbied hard for its enactment) which gave protection to the engraver for a period of fourteen years.[146] Under its terms, if a map was copied illegally the print and copperplate could be confiscated and destroyed and a fine of five shillings levied for each print made. The Act required that the engraver's name, 'the Name and Address of the Proprietor', should be printed on every map, together with the 'Date of First Publishing'. Further Acts in 1767 and 1777 extended the duration of copyright to twenty-eight years. In reality, very few cases of copyright ever came to court as a result of these Acts. Legal costs were high, and if a few changes had been made to the plate it could be argued successfully that the copy was an improvement on the original, and thus of value to the populace. Moreover, while awaiting the court case (which could be delayed by as much as six years) sale was prohibited not only of the disputed copy, but also of the original on which it was based. There was thus little sense in taking such a matter to law, and from the manner in which engraving plates passed freely within the close circle of map-makers it is clear that the Act was generally ignored.

Financing the county maps

Publishing county maps was a costly undertaking and several map-makers were impoverished by their involvement. William Day described how his map of Somerset '… cost me more than a thousand guineas, exclusive my great labour and trouble … we were more than Seven years constant employ surveying it'.[147] Donn's map of Devon cost almost £2,000 to produce,[148] while Yeakell and Gardner estimated that £2,400 would be required over six years for surveying, drawing and engraving their map of Sussex.[149] Faden's Norfolk map had already cost him £2,000 by 1794, the last year of the survey – and still three years before publication. Andrew Armstrong, in his correspondence with the Society of Arts, listed the expenses incurred during his survey of Northumberland, which was carried out between September 1766 and March 1769:[150]

Myself, one assistant, 2 labourers (sometimes more), horses	£350 0 0
9 copper plates (24 by 18 ins.)	£ 14 10 0
Engraving	£ 90 0 0
Paper for 500 maps	£ 27 0 0
Copper plate printing	£ 20 0 0
Printing the Index 500 copies	£ 10 0 0
Printing advertisements and proposals	£ 5 0 0
	£516 10 0

Henry Beighton stated that his Warwickshire map of 1728 would cost between £30 and £40 pounds to engrave. Even when all surveying and printing were completed, costs could continue to mount up. William Norris, bookseller of Taunton, wanted to take a quarter of the subscriptions he collected for Strachey's map of Somerset as his fee, while Lindley was charged £1 16s for insuring the plates of his Surrey map for four years.[151] These figures suggest that even the best prize of £100 from the Society of Arts would not have gone very far in covering the costs of producing and publishing the county maps. Thomas Jefferys was finally made bankrupt by his county map venture, and others were quick to sell their copperplates to cover debts.[152] Those for Burdett's map of Cheshire were sold in Mr Christie's auction rooms in Pall Mall in 1784 and eventually came into the hands of Faden.[153] Ever the canny business man, he also bought the plates of Chapman's Nottinghamshire map for forty-five guineas in 1784, and of Yates's Staffordshire survey, for £95, in 1789 (the Society of Arts offered Faden the choice of either a gold medal or £50 in cash as a prize for his 1791 map of Hampshire: he characteristically 'chose the pecuniary award').[154]

The problem for the county mapmakers was that funds were needed for the survey several years before they could expect to see any financial return on their investment. The usual solution was to seek subscriptions from the local gentry well in advance of publication and even before surveying had begun. In return for their 'encouragement' the names of the participating individuals, 'engraved at their respective seats', were given prominence on the maps, and on Warburton's maps their coats of arms were displayed as a decorative border (for a price of 12s 6d a time). Similarly, the proposals for Jefferys' map of Huntingdon note that, for an extra half-guinea, the family arms would appear on the map, while 'those freeholders who chuse to have their names inserted to their seat' were invited to give notice to the publisher: Beighton made the same offer regarding his Warwickshire map.[155] Subscriptions were sought through advertisements in newspapers and they came almost invariably from people living within the county, occasionally from an adjacent county, but rarely from further afield: George Adams, an instrument-maker in London who subscribed to Donn's map of Devon, was one of the exceptions, although others came from Bristol, Bath and Oxford. Most subscribers were members of the landowning gentry or clergy, although in the case of Jefferys' map of Yorkshire the names of industrialists are also prominent.[156] Such financial support was vital: Peter Burdett stated that he needed 400 subscribers before he would even consider surveying Lancashire. But he also complained that 'two-thirds of the Subscribers omitted to pay in their first Subscription of One Guinea ... and desire they will without delay pay'.[157]

Some maps were dedicated to a particularly notable member of the nobility – the Duke of Grafton of Euston Hall in the case of Hodskinson's map of Suffolk – but it is unclear whether this was in return for a particularly large additional payment. Peter Burdett's map of Derbyshire was unusual in that it was dedicated 'to the President, Vice President and the rest of the members of the Society for the Encouragement of Arts': it may, or may not, have been a coincidence that he was awarded one of the Society's £100 prizes! Occasionally a map was financed privately, as in the case of Dickinson's map of South Yorkshire, 'taken at the cost of the most Hon. Thomas Marquess of Rockingham'.[158] For his Staffordshire map, Yates omitted the names of estate owners from his map; this may well have been a commercial mistake, because when the copperplate eventually passed into the hands of William Faden 158 names were added for a new edition.[159] William Yates surveyed Lancashire at his own expense, probably because Peter Burdett failed to return the subscriptions he had received for his earlier, abortive, attempt at mapping the county. In Norfolk, as we shall see, Mostyn Armstrong's similar failure ensured that members of the gentry were unwilling to pay any money to William Faden for making his map of the county until the survey was actually completed.

The list of subscribers, which was sometimes published at the same time as the map, is a gold mine for social historians, including as it does more names than could possibly have fitted onto the map itself – for obvious reasons, town-dwellers are considerably under-represented. It is a feature that defines the age in which the maps were published; a similar exercise today would be almost inconceivable. But even when – as in the case of Norfolk – the list has not survived, the names on the map itself represent an important insight into contemporary social geography. The proposals for Yates's Lancashire map make it clear that the names of all subscribers will appear on the map; in the case of Cumberland, however, there is a suggestion that this privilege could be secured only by making an additional payment. The evidence for Norfolk in this regard is unfortunately missing. However, Hodskinson's Suffolk map bears the names of 150 individuals, while those for Cumberland and Westmorland bear 96 and 38 respectively.[160] The subscription list for John Prior's map of Leicestershire includes 264 names[161] and that for Chapman and Andre's Essex, 240,[162] while William Yates's Lancashire has 252 and Donn's Devonshire no less than 528.[163] The number of subscribers may give us an idea, but no more than that, of how many maps were sold. Evans's map of north Wales had 280 subscribers, but they collectively ordered 658 copies,[164] and it is clear from the Lindley papers that extra copies of his Surrey map, beyond those which had been subscribed for, were ordered by the publishers. Lindley, in a legal battle with his partner's executors, stated that he had personally delivered 575 copies of the map to William Faden.[165] The orders made by both the Duke of Queensberry and the Earl of Shelburne for forty copies of the 1773 map of Wiltshire (at a cost of two guineas a set) were exceptional,[166] but even these figures do not compare with the 100 copies of Donn's map of Devonshire which were ordered by John Baring of Mount Radford.[167] Henry Beighton tried to put pressure on potential purchasers of his Warwickshire map by stating that 'no more shall be printed than what are subscribed for'.[168] This technique does not appear to have been very successful, however, as only about 240 maps were thus paid for in advance: he clearly relented, and after the initial publication he added further family coats of arms as fresh orders came through. There is some evidence that arms were, on occasion, removed from the copperplate before second or third printings, perhaps because particular families had failed to come up with the promised payments. In summary, we simply do not know how many copies of any of the county maps were printed, or how many were sold. A very rough guess would give us a figure of between 200 and 800.

We do not know how many copies of Faden's Norfolk map were printed: nor for that matter do we have any clear idea of how many have survived to this day. Some are to be found in county record offices and libraries but the majority are likely to be in private hands, some still in the libraries of the country houses which are illustrated on the map.[169] Because of his important connections with

America, with which we shall deal in the next chapter, many are in public and university libraries in the United States and Canada.

The price of the maps

The Norfolk map, at two and a half guineas to subscribers and three to non-subscribers, was expensive when compared with other county maps of the period, although its physical size may have been one reason for this. Jefferys' 1768 survey of Huntingdonshire, for example, cost only one guinea, half of which was payable in advance.[170] The Liverpool *General Advertiser* of 30 November 1786 stated that William Yates's Lancashire map would be on sale to subscribers in eight sheets at one and a half guineas, or 'pasted on canvass, and neatly coloured at two pounds seven shillings'.[171] Commercially viable colour-printing was not available until the late nineteenth century so this embellishment would have been undertaken by hand, probably at a cost of around five shillings per map. Benjamin Donn's 1765 Devon map sold for one and a half guineas, but the sheets were sometimes bound together with a title page, index map and a printed index, and sold for two guineas, and a limited number were printed on vellum.[172] Peter Burdett's map of Cheshire was printed on paper in four sheets and sold for two guineas or, at an unspecified price, on linen, dissected and folded. The same map was also available mounted on canvas and rollers at a price of £2 11s 6d.[173] (The county maps have generally survived best in dissected and folded form, and for understandable reasons are rarely found in reasonable condition in the roller version.) The price of maps depended in part also on market conditions. The maps generally sold well immediately after publication, but sales soon fell away; Lindley wrote of his Surrey survey that 'the map wo'd have made us a great deal of money … but the Business is become Stale'.[174] Day and Master's map of Somerset sold at £1 11s 6d in nine sheets when first published in 1782, but six years later this was reduced to a guinea, or £2 7s 0d for a version 'elegantly coloured and fitted up on canvas and rollers', perhaps to boost flagging sales.[175] Unusually, and possibly as a salesman's gimmick, the maps of Derbyshire by Peter Burdett 'are numbered and signed by Mr. Snowden [who had acquired the copperplate]'. The cost of many maps was drastically reduced in the early nineteenth century, when the new surveys by Bryant and the Greenwoods appeared on the horizon (although the Somerset map, by then in the hands of William Faden, was holding its price at £1 12s as late as 1822).[176] By this date the cost of his 1797 Norfolk map had been reduced to two guineas, or ten shillings for the reduced version on one sheet – this was only four years before Bryant's new survey of Norfolk was published.

The contrast with France

Up until the last quarter of the eighteenth century the pre-eminence of French cartography was a constant thorn in the side of the London map-makers, academics and politicians. England simply did not have the system of government supported scholar-geographers that had been the norm in France for many years.[177] In addition, entry into the engraving and printing trade was easier in France, where the trade guilds were less restrictive.[178] A trigonometrically based survey of the entire country, the *Carte de France*, was initiated in 1747 by Louis XV and the Cassini de Thury family (four generations were involved in succession at the centre of the French map trade): this was almost half a century before the Board of Ordnance was established in England. The whole country was eventually mapped in 173 sheets at a scale of 1:86,400. The project was initially given financial support by the king, and although the Seven Years War dampened royal enthusiasm for the project, the services of military surveyors continued to be made available free of charge, and the family were granted a thirty-year monopoly on sales. Private subscriptions were sought at times but for the most part the endeavour, which cost around £2,000 (40,000 livres) per sheet including surveying, engraving and printing, was supported by the various provincial governments.[179] The contrast with the situation in Britain, where the series of county maps was produced entirely by private business and the government's involvement was marked by its absence, is striking. In a sense this was a reflection of the generally more decentralised character of political power in Britain, 'widely and almost randomly distributed among a bewildering array of agencies and individuals, including Crown, ministry, both houses of Parliament, Church of England etc.'[180] The only indirect financial support for the county maps provided by the Crown came when George III offered to pay for Jesse Ramsden's theodolite in the 1780s.[181] A further contrast with the situation in France was that the drafts of county maps were sent for the correction of names and minor topography not to the nobility, but to the local parish priest.[182] The *Carte de France* was taken over by the Dépôt de la Guerre in 1793 and, for military reasons, no maps were published for several years after this. But by 1818 the French project was complete. William Mudge, the first director of the Trigonometrical Survey in England, called the *Carte de France* 'a performance highly celebrated'.

Notes

1   John Green, whose real name was Braddock Mead (1680–1757), worked with map-maker Thomas Jefferys, William Faden's predecessor. G. R. Crone, 'Further notes on Braddock Mead, alias John Green', *Imago Mundi* 8 (1951), p. 69.
2   P. Laxton, 'The geodetic and topographical evalution of English county maps 1740–1840', *Cartographic Journal* 13 (1976), pp. 37–54.
3   C. Delano-Smith and R. J. P. Kain, *English Maps: A History*. The British Library, London, 1999, p. 66.
4   W. L. D. Ravenhill, 'Introduction', in *Christopher Saxton's 16th Century Maps: The Counties of England and Wales*. Chatsworth Library, Shrewsbury, 1992, p. 23.
5   R. Frostick, *The Printed Plans of Norwich 1558–1840: A Carto-bibliography*. Privately published, Norwich, 2002, p. 1.
6   Delano-Smith and Kain, *English Maps*, p. 69.

7   J. B. Harley, 'Christopher Saxton and the first atlas of England and Wales 1579', *The Map Collector* 8 (1979) pp. 3–11.

8   J. B. Harley, 'From Saxton to Speed', *Cheshire Round* 1 (1966) pp. 174–84.

9   R. A. Skelton, *County Atlases of the British Isles 1579–1850: Bibliography 1579–1703*. Carta Press (1970), pp. 124, 125, 149.

10 R. Gough, *British Topography*, vol. 1. London, 1780, Preface, xvi. Richard Gough (1735–1809), antiquarian, has been described by J. B. Harley as the father of the systematic study of early English maps. In his two volumes of *British Topography* he catalogued the maps and charts then available for study. His writings had considerable influence on map-making and map-collecting. Sir George Cullum (1733–1815), antiquary, said of Gough's *Topography*: 'next to the Bible it is the book I most often read' (J. B. Harley and G. Walters, 'English map collecting 1790–1840; a pilot survey of the evidence in Sotheby sale catalogues', *Imago Mundi* 2nd Series 30, 4 (1978), pp. 31–55, at p. 34).

11 Delano-Smith and Kain, *English Maps*, pp. 78, 105 and 175. Also P. Hindle, *Maps for Historians* Phillimore, London, 1998, p. 94.

12 J. B. Harley, 'The re-mapping of England 1750–1800', *Imago Mundi* 19 (1965), pp. 56–67.

13 W. L. D. Ravenhill and O. J. Padel, *A Map of the County of Cornwall newly surveyed by Joel Gascoyne*. Devon and Cornwall Record Society New Series 34, Plymouth, 1991, pp. i–x and 1–37, at p. 9.

14 E. M. Rodger, *The Large Scale County Maps of the British Isles 1596–1850: A Union List*, 2nd edn. Bodleian Library, Oxford, 1972.

15 P. D. A. Harvey and H. Thorpe, *The Printed Maps of Warwickshire 1576–1900*. Warwick County Council in association with the University of Birmingham, 1959, pp. 19–35.

16 Gough, *British Topography*, vol. 2, p. 297, and RSA, *Manuscript Transactions*, 1782–3, p. 10.

17 R. Frostick, 'James Corbridge – Surveyor', *Norfolk Archaeology* 45, 2 (2007), pp. 155–75.

18 D. Stoker, 'An eighteenth century map piracy', *Norfolk Archaeology* 37, 1 (1979), pp. 123–6.

19 A trigonometrical survey on the county was carried out by Dr Charles Mason, of Trinity College and Professor of Geology at the university. He died in 1759 before the topographical part of the map was complete. J. B. Harley, 'The Society of Arts and the surveys of the English counties 1759–1809', *Journal of the Royal Society of Arts* 112 (1963–64), p. 123.

20 LRO, DD He 61/22: P. P. Burdett, *The Proposals for P. P. Burdett's Map of Lancashire 1768*.

21 C. Close, *The Early Years of the Ordnance Survey*. Repr. David and Charles, Newton Abbot, 1969 (originally published 1926), p. 30.

22 E.g. W. Mudge, *An Account of the Operation for Accomplishing a Trigonometrical Survey of England and Wales, 1797–1799*, vol. 2. London, 1801; can be downloaded in full from www.books. google.com.

23 W. Mudge and I. Dalby, 'An account of the operations carried on for accomplishing a trigonometrical survey of England and Wales 1784–1796', *Philosophical Transactions* 1 (1799), p. 411.

24 D. Kingsley, *Printed maps of Sussex 1575–1900*. Sussex Record Office 72, 1980–81, p. 108.

25 R. A. Skelton, *Saxton's survey of England and Wales*. Nico Israel, Amsterdam, 1974, pp. 7–8.

26   P. Langford, *A Polite and Commercial People. England 1727–1783*. Oxford, Oxford University Press, 1992; S. Tarlow, *The Archaeology of Improvement in Britain, 1750–1850*. Cambridge University Press, Cambridge, 2007; A. D. M. Phillips, 'Introduction', in *A map of the county of Stafford by William Yates, 1775*. Staffordshire Records Society 4th series, vol. 12, Stafford, 1983, pp. ii and iii.

27   RSA, *Guard Books 2*, no. 85, 11 February 1756. The Royal Society of Arts library is housed at 8, John Adams Street, London. The archives are in the form of Guard Books, Committee Minutes and Transactions.

28   H. T. Wood, 'The Royal Society of Arts: the Premiums. History of the premiums given for large scale county maps 1759–1809', *Journal of the Royal Society of Arts* 60 (1912), pp. 263–74.

29   Harley, 'The Society of Arts'. Borlase's letter is in the Manchester University John Ryland's Library, English MS, 19, VI, 178.

30   RSA, *Minutes of Committee (Polite Arts)*, 10 March, 1759.

31   Harley, 'The Society of Arts', p. 44.

32   RSA, *Minutes of Committee (Polite Arts)*, 10 March 1760.

33   RSA, *Guard Book*, IV, 116.

34   Gough, *British Topography*, vols 1 and 2.

35   RSA, *Minutes of Committee (Polite Arts)*, 5 June 1767.

36   Harley, 'The re-mapping of England', p. 60. This view is not universally shared. Laurence Worms (pers. comm.) feels that the influence of the Royal Society has been heavily overstated. He feels that the work was already well underway before the offer of prizes and their monetary value was minimal in the context of the overall cost of survey and engraving. He thinks the maps would have been produced in any event. A contemporary had similar thoughts; Steven McDougall, a Glasgow estate surveyor in 1760, said the 'premium was too little for a man to execute the survey' and hinted that 150 guineas would be needed to tempt him south to survey Somerset. RSA, *Guard Books*, IV, 105.

37   Harley's papers and books are listed in the bibliography. His analyses of the county maps of Surrey, Lancashire, Somerset, Yorkshire and Derbyshire are among the most detailed that have been published.

38   G. Walters, 'Themes in the large scale mapping of Wales in the eighteenth century', *Cartographic Journal* 5 (1968), p. 140.

39   *Ibid.*, p. 140.

40   C. Withers, *Placing the Enlightenment: Thinking Geographically about the Age of Reason*. University of Chicago Press, Chicago, 2007, pp. 202–4 and 224.

41   M. S. Pedley, *The Commerce of Cartography: Making and Marketing Maps in Eighteenth Century France and England*. University of Chicago Press, Chicago, 2005, p. 1.

42   In the 18th century the word geographer had a more limited meaning; it meant essentially a map-maker, engraver or publisher. The definition in Thomas Sheridan's *Dictionary of the English Language*, 2nd edition (1789) was 'one who describes the earth according to the position of its different part'. The word cartographer did not appear in the English language until 1859 (Oxford Universal English Dictionary 1937) and 'cartographe' in French in 1877. See Pedley, *Commerce and Cartography*, note 2, p. 248.

43   Pedley, *Commerce of Cartography*, p. 30.

44   *Ibid.*, p. 5.

45   P. Eden (ed.), *Dictionary of Land Surveyors and Local Map Makers of Great Britain and Ireland 1530–1850*, 2 vols, 2nd edn by Sarah Bendall. British Library, London, 1997.

46   Pedley, *Commerce of Cartography*, p. 165.

47   W. A. Seymour (ed.), *A History of the Ordnance Survey*. William Dawson, Folkestone, 1980, p. 49.

48   R. A. Skelton, *Decorative Printed Maps of the 15th to 18th Centuries*, Staples Press, London, 1952, p. 1.

49   Robert Walton, *A New Booke of Mapps*, 1655, 71–72, quoted by S. Tyacke, 'Map-sellers and the London map trade 1650–1710', in *My Head is a Map: Essays and Memoirs in Honour of R. V. Tooley*. Carta Press, London, 1973, p. 66.

50   Gough, *British Topography* (BLO, shelfmark Gough, GEN, Top 363–6).

51   Pedley, *Commerce of Cartography*, p. 89.

52   Harley and Walters, 'English map collecting', p. 48.

53   J. B. Harley and P. Laxton, *Peter Burdett's Map of Cheshire, 1777: The Theory and Practice of an Eighteenth-Century County Survey*. Historic Society of Lancashire and Cheshire, Occasional series Vol. 1. Published with a facsimile and introduction by Lund Humphries, 1974, p. 6; Pedley, *Commerce and Cartography*, p. 47.

54   J. Strawhorn, 'Introduction', in *A Map of Ayrshire 1775 by A. and M.J. Armstrong*. Collections of the Ayrshire Archaeology and Natural History Society, 1959, p. 244.

55   J. B. Harley, 'Introduction', in William Yates, *A Map of the County of Lancashire 1786*. Historic Society of Lancashire and Cheshire, Birkenhead, 1968.

56   J. B. Harley, 'John Strachey of Somerset: an antiquarian cartographer of the early eighteenth century', *Cartographic Journal* 3, 1 (1966), pp. 2–7.

57   W. L. D. Ravenhill, 'Introduction', in *Benjamin Donn's map of the county of Devon, 1765*. Devon and Cornwall Record Society New Series 9, University of Exeter, 1965, pp. 1–20. p. 1.

58   W. L. D. Ravenhill, 'Benjamin Donn 1729–98: map-maker and master of mechanics', *Transactions of the Devonshire Association* 97 (1965), pp. 179–93 at p. 181.

59   J. D. Welding, 'Introduction', in *Leicestershire in 1777: An Edition of John Prior's Map of Leicestershire*. Leicestershire Libraries and Information Services, 1984.

60   *Ibid.*

61   J. B. Harley, 'English county map-making in the early years of the Ordnance Survey: the map of Surrey by Joseph Lindley and William Crosley', *Geographical Journal* 132, 3 (1966), pp. 372.

62   M. Ravenhill and M. Rowe, *Devon Maps and Map-Makers: Manuscripts and Maps before 1840*. Devon and Cornwall Record Society New Series 43, Exeter, 2002, pp. 1–25, at p. 17.

63   Harvey and Thorpe, *Printed Maps*, p. 22.

64   A. H. W. Robinson, *Marine Cartography in Britain*. Leicester University Press, Leicester, 1972, p. 74.

65   Pedley, *Commerce of Cartography*, p. 22.

66   Withers, *Placing the Enlightenment*, p. 105.

67   I. Evans and H. Lawrence, *Christopher Saxton: Elizabethan Map-Maker*. Wakefield Historical Publications, Holland Press, 1979, p. 80.

68   L. Worms, 'Location in the London map trade', *International Map Collectors' Society Journal* 28, (2000), pp. 32–42.

69   E. Crittall, 'Introduction', in *Andrews' and Dury's 1773 Map of Wiltshire*. Wiltshire Archaeological and Natural History Society 8, 1952.

70   Delano-Smith and Kain, *English Maps*, p. 156.

71   D. Sobel, *Longitude*. Fourth Estate, London, 1996, pp. 11–16; Delano-Smith and Kain, *English Maps*, p. 78.

72   J. Ainslie, *A Comprehensive Treatise on Land Surveying Comprising the Theory and*

Practice in all its Branches. Edinburgh, 1812, p. 219. This edition is available in full from www.books.google.com. The book is dedicated to John Rennie, a Smeatonian like William Faden. Its fifth chapter is devoted to county surveying. The other book widely used was A. Burns's *Geodaesia Improved or a New and Correct Method of Surveying.* Tarporley, Chester, 1771. This is also available from www.books.google.com.

73   G. Adams, *Geometrical and Graphical Essays containing a General Description of the Mathematical Instruments used in Geometry, Civil and Military Surveying, Levelling and Perspective.* London, 1791 (the 4th edn of 1813, corrected by William Jones, is available in full from www.books.google.com).

74   A. W. Richeson, *English Land Measuring to 1800: Instruments and Practice.* Society for the History of Technology, Massachusetts Institute of Technology, 1966, p. 109.

75   J. B. Harley, 'William Yates and Peter Burdett: their role in the mapping of Lancashire and Cheshire during the late eighteenth century', *Transactions of the Historic Society of Lancashire and Cheshire* 115 (1964), p. 110.

76   Harley and Laxton, *Burdett's Map of Cheshire*, p. 14.

77   *Ibid.*, p. 15.

78   Withers, *Placing the Enlightenment*, p. 97.

79   W. L. D. Ravenhill, 'Introduction' in *Two Hundred and Fifty Years of Map-making in the County of Surrey: A Collection of Reproductions of Printed Maps Published Between the Years 1579–1823.* Harry Margary, Lympne Castle, 1974.

80   L. Worms, 'John Seller', in *The Dictionary of National Biography.* Oxford University Press, Oxford, 2004.

81   J. C. Barringer, 'Introduction', in *Faden's Map of Norfolk.* Larks Press, Dereham, 1989, p. 4.

82   Ravenhill, 'Introduction', 1974.

83   Harley and Laxton, *Burdett's Map of Cheshire*, p. 6. Above the triangulation diagram on the Cheshire map Burdett said that he 'submitted the map to the Inspection, and intended for the use of the curious in Geography'.

84   Ainslie, *Comprehensive Treatise*, p. 220.

85   Harley and Laxton, *Burdett's Map of Cheshire*, p. 17.

86   Ainslie, *Comprehensive Treatise*, p. 219.

87   Harley, 'English county map-making', p. 373.

88   Harvey and Thorpe, *Printed Maps*, p. 35.

89   J. B. Harley and R. W. Dunning, 'Introduction', in *Somerset Maps: Day and Masters 1782 and Greenwood 1822.* Somerset Record Society 76, Taunton, 1981.

90   Harley, 'English county map-making', p. 375.

91   Ainslie, *Comprehensive Treatise*, p. 221.

92   Adams, *Geometrical and Graphical Essays*, p. 206.

93   Richeson, *English Land Measuring*, p. 164.

94   Richeson, *English Land Measuring*, pp. 172, 180.

95   *Ibid.*, pp. 163, 172 and 198.

96   Adams, *Geometrical and Graphical Essays* (1813 edition), p. 532.

97   Welding, 'Introduction', p. 7.

98   Office for National Statistics, *1801 Census.*

99   M. Turner, *English Parliamentary Enclosure: Its Historical Geography and Economic History.* W. Dawson, Folkestone, 1980, pp. 70, 78.

100   A. Macnair, ' Foulsham and North Elmham: the landscape of two Norfolk parishes at the time of parliamentary enclosure', *Norfolk Archaeology* 44, 2 (2003), pp. 269–92.

101   A fourth variable could perhaps be added – the veracity of the recording of the

commons on any individual county map. Harley has shown that in Burdett's map of Derbyshire (which won a Society of Arts prize) some commons are named but no attempt is made to indicate their perimeter and several commons are not noted at all. J. B. Harley, D. V. Fowkes and J. C. Harvey, 'Introduction', in *Burdett's Map of Derbyshire, 1791*. Derbyshire Archaeological Society, Derby, 1975.

102   Harley, 'John Strachey', p. 5.

103   Harley, 'Introduction', p. 14.

104   Laxton, 'Geodetic and topographical evaluation', p. 44.

105   R. A. Skelton, 'The origins of the Ordnance Survey of Great Britain', *Geographic Journal* 128 (1962), p. 417.

106   P. Laxton 'Introduction', in *County of Berkshire Map by John Rocque*. Harry Margary, Lympne Castle, 1972.

107   Harley and Harvey, 'Introductory notes', p. 3.

108   On Joel Gascoyne's map of Cornwall the 'pricked [or dotted] roads are open roads upon Downs or commons and [undotted] roads are inclosed'.

109   P. Hindle, 'The first large scale county maps of Cumberland and Westmorland in the 1770s', *Transactions of the Cumberland and Westmorland Antiquarian & Archaeological Society* 3rd Series 1 (2001), pp. 139–53.

110   Crittall, 'Introduction'.

111   Hans Gyger, a Swiss mathematician, surveyor and cartographer in 1664 produced a map of Zurich using shaded relief for the first time. The original map can be seen in the Museum Haus zum Rechberg, Zurich.

112   Ordnance Survey first edition six-inch maps.

113   Seymour, *Ordnance Survey*, p. 126.

114   Hutton, C. 'An account of the calculations made from the survey and measures taken at Schiehallion, in order to ascertain the mean density of the Earth', *Philosophical Transactions of the Royal Society* 60 (1778), 689–788. For details of Maskelyn's mapping interests see R. V. Tooley, *Dictionary of Mapmakers*, revised edition by Josephine French. Map Collector Publications, London, 1999.

115   Harley, 'Introduction', p. 18.

116   With this proviso it is worth noting that Jefferys's *Yorkshire* has 736 watermills (175 on the River Don and its tributaries) and 161 windmills, Burdett's *Cheshire* has 16 windmills and 156 watermills, Yates's *Lancashire* has 330 watermills and 79 windmills, Donald's *Cumberland* has 141 watermills and 3 windmills, Donn's *Devon* has over 100 watermills and 5 windmills and Prior's *Leicestershire* has 92 watermills and 81 windmills.

117   Harley and Harvey, 'Introductory notes'.

118   *Ibid.*

119   Harley and Laxton, *Burdett's Map of Cheshire*, p. 33.

120   Harley *et al.*, 'Introduction'.

121   Harvey and Thorpe, *Printed Maps*, p. 30.

122   Seymour, *Ordnance Survey*, p. 70.

123   Pedley, *Commerce of Cartography*, p. 45.

124   D. Woodward (ed.), *Five Centuries of Map Printing*. University of Chicago Press, Chicago, 1975, p. 64. The third chapter, on copperplate printing, is by Coolie Verner.

125   Pedley, *Commerce of Cartography*, p. 44.

126   *Ibid.*, p. 43.

127   Harley and Dunning, 'Introduction', p. 18.

128   Harley and Laxton, *Burdett's Map of Cheshire*, p. 2.

129 Harley, 'Introduction'. Harley states that William Yates's maps 'mark the real culmination of pre-Ordnance Survey cartography'.

130 P. Laxton, 'Introduction', in *Two Hundred and Fifty Years of Map-making in the County of Hampshire: A Collection of Reproductions of Printed Maps Published Between the Years 1575 and 1826*. Harry Margary, Lympne Castle, 1976.

131 Strawhorn, 'Introduction', p. 244.

132 Woodward, *Map Printing*, p. 76.

133 Harley and Harvey, 'Introductory notes'.

134 Woodward, *Map Printing*, p. 69.

135 Pedley, *Commerce of Cartography*, p. 64.

136 Woodward, *Map Printing*, pp. 67, 68, quoting William Faithorne, *The Art of Graveing and Etching*, 2nd edn. London, 1702.

137 Seymour, *Ordnance Survey*, p. 70.

138 Harley and Laxton, *Burdett's Map of Cheshire*, p. 8.

139 J. Balston, *The Elder James Whatman 1702–1759: England's Greatest Paper-Maker*. Privately published, West Farleigh, Kent, 1992; and see also www.wovepaper.co.uk (accessed 30 June 2009).

140 RSA, *Minutes of Committee (Polite Arts)*, 5 June 1767; E. J. Labarre *Dictionary and Encyclopaedia of Paper and Paper-making*, 2nd edn. Amsterdam, 1952, p. 332.

141 Pedley, *Commerce of Cartography*, p. 65.

142 Phillips, 'Introduction', p. xxix.

143 Harvey and Thorpe, *Printed Maps*, p. 35. For his Warwickshire map Henry Beighton devised a scale to compensate 'for ye Swelling and Shrinking of Paper in Hott and Hazy weather'.

144 Woodward, *Map Printing*, p. 66.

145 The Statute of Anne, 1709 'An Act for the Encouragement of Learning, by vesting the Copies of Printed Books in the Authors or purchasers of such Copies, during the Times therein mentioned'.

146 Pedley, *Commerce of Cartography*, p. 102.

147 RSA, *Minutes of Committee (Polite Arts)*, 20 December 1782.

148 Ravenhill, W. L. D. 'Introduction', in *Benjamin Donn's map of the county of Devon, 1765*. Devon and Cornwall Record Society New Series 9, University of Exeter, 1965, p. 6.

149 On the note of publication of the first sheet of the map in 1778 inserted in Richard Gough's collection for a 3rd edition of his *British Topography*.

150 Harley, 'The Society of Arts', p. 269.

151 Harley, 'Strachey's Map of Somerset', p. 5; Harley, 'Map of Surrey', pp. 372–378.

152 J. B. Harley, 'The bankruptcy of Thomas Jefferys: an episode in the economic history of eighteenth-century map-making', *Imago Mundi* 20, (1966), pp. 44.

153 Harley, 'Re-mapping', p. 67. The *Derby Mercury* for 23 December 1784 records that the copper plates of Burdett's Derbyshire 'were sold out last week by auction at Mr. Christies in Pall Mall'.

154 Laxton, 'Introduction', p. 5; Harley, 'The Society of Arts', p. 273; RSA, *Society Minutes*, 1 May 1793.

155 Harvey and Thorpe, *Printed Maps*, p. 23.

156 Harley and Harvey, 'Introductory notes'.

157 *Manchester Mercury*, 5, 12 and 19 February 1771.

158 Gough, *British Topography*, vol. 2, p. 477.

159 Phillips, 'Introduction', p. xxv.

160 Hindle, 'Cumberland and Westmorland'. Contains a full list of subscribers.

161 Welding, 'Introduction'.

162   *Essex Review*, Vol. 19. (1910), pp. 83–88.

163   Ravenhill 'Introduction', *Benjamin Donn's Map of Devon*, p. 7.

164   Harley, 'The Society of Arts', p. 274.

165   Harley, 'English county map-making', p. 375.

166   Crittall, 'Introduction'.

167   Ravenhill 'Introduction', *Benjamin Donn's Map of Devon*, p.7.

168   Harvey and Thorpe, *Printed Maps*, p. 24.

169   E. Rees and G. Walters, 'The library of Thomas Pennant', *The Library, Transactions of the Bibliographical Society* 5th Series 25, 2 (1970), p. 147, quotes Thomas Pennant as writing to a correspondent in 1793: 'I must beg to say I have Armstrong's great map of Northumberland, and the same of Durham, and the enormous map of Yorkshire.'

170   Harley, 'Bankruptcy of Thomas Jefferys', p. 43.

171   Harley, 'Introduction', p. 14.

172   Ravenhill 'Introduction', *Benjamin Donn's Map of Devon*, Plate 2.

173   Harley and Laxton, *Burdett's Map of Cheshire*, p. 8.

174   YAS, MD/280/1, quoted in Harley, 'English county map-making', p. 376.

175   *Western Flying Post*, 7 July 1788, quoted in Harley and Dunning, 'Introduction', p. 28.

176   W. Faden, *Catalogue of the geographical works, maps, plans &c.* Map Collectors Circle, London, 1963.

177   Pedley, *Commerce of Cartography*, p. 199.

178   *bid.*, p. 200.

179   *Ibid.*, p. 70.

180   W. Prest, *Albion Ascendant: English History 1660–1815*. Oxford University Press, Oxford, 1998, p. 157.

181   Pedley, *Commerce of Cartography*, p. 83.

182   *Ibid.*, p. 186.

CHAPTER 2

Faden's Map of Norfolk

Geographie is the imitation, and description of the face, and picture of th'earth.

William Cunningham, 1559[1]

Introduction

The map of Norfolk published on 12 August 1797, the subject of this short book, is usually named after its publisher, William Faden. It might more reasonably be named after its two surveyors, Thomas Donald and Thomas Milne. Why Faden's name should have thus become immortalised is uncertain: the map of Suffolk published in 1783 has always been universally named after its surveyor, Joseph Hodskinson, even though it was likewise published by Faden.

Norfolk is fortunate in that its first large-scale map was produced by two of the most able surveyors of the late eighteenth century, working in association with the most dynamic and competent map publisher of his period. Moreover, it was published very late in the series of county maps and thus benefited from the lessons learnt by its makers from their earlier endeavours. Things could quite easily have turned out very differently if Mostyn Armstrong had been successful in *his* attempts to produce a county map some two decades earlier: this would almost certainly have been much inferior in quality.

No history of the Norfolk map would be complete without a brief note on Thomas Jefferys (1719–1771), whose business was taken over by William Faden: for the latter's success owed much to his predecessor. Jefferys was the son of a cutler and was, like Faden, originally apprenticed as an engraver. The first part of his career was spent producing maps of manufacturing towns in the Midlands, and it was the Seven Years War (1756–1763) that gave him the opportunity to publish a far greater range of maps, for the theatres of war in this Anglo-French struggle included much of Europe, North America and India. By publishing such things as Joshua Fry and Peter Jefferson's *A Map of the most Inhabited part of Virginia* and, in association with Robert Sayer in 1768, the atlas *A General Topography of North America and the West Indies*, Jefferys rapidly became a leading expert on maps of foreign, and especially American, places. This was recognised in 1760 when he became Geographer in Ordinary to George III (although he had previously been Geographer to the Prince of Wales since at least 1746). He set up his home and business premises, which for the next century was to be the leading London map retailing establishment, on the south-east corner of St Martin's Lane and the Strand.[2] When the war ended he turned his attentions closer to home, and to the production of English county maps. His initial foray into this field simply involved engraving and printing Benjamin Donn's 1765 map of Devon, which (as we have seen) won the first Society of Arts prize. He soon decided to become involved in such maps at an earlier stage in their production, and thus initiated new surveys of Bedfordshire in 1765,[3] Huntingdonshire, Oxfordshire and Buckinghamshire in 1766, and the Holland district of Lincolnshire in 1767. He employed three of the best surveyors available, John Ainslie (his own apprentice), Joseph Hodskinson and Thomas Donald. Unfortunately he overstretched himself financially, and was made bankrupt in November 1766.[4] Nevertheless, with the help of friends – especially Robert Sayer, a successful map and print dealer – he continued to publish county maps, and was involved in the production of maps of Westmorland and Cumberland (with Thomas Donald again working as surveyor in both cases), Yorkshire, Northamptonshire and Durham. Despite their very high standard, none of his maps won a Society of Arts prize. When he entered his Oxfordshire map for consideration the Society ordered that 'his letter be laid aside as he did not pay his arrears at the time he declined (i.e. left the society) nor since; and that the map be returned to him': in other words, he was unable to keep up his subscriptions because of his financial problems.[5] In retrospect this seems a somewhat petty reason to ignore his significant contribution to English cartography. As well as producing maps of counties, Jeffreys also published a number connected with industrial and commercial projects, such as John Smeaton's *A Plan of the Harbour of King's Lynn* (1767) and John Grundy's *A Plan of the intended Canal from Chesterfield* (1770). He died in 1771,[6] leaving only £20 to his wife Elizabeth, together with a large collection of maps and charts, many of which were sold in 1772.[7] In his prime he had been a very successful cartographer: it was the expense involved in surveying the English counties that finally ruined him, a warning to William Faden, who took over his business and shop at the age of 24.

William Faden was born in 1749 off Fleet Street in London, just across the road from Robert Sayer who, as we have seen, was later to be so helpful to Thomas Jefferys. His father, also called William, was named MackFaden for the first thirty years of his life; but, like many other Scotsmen, he changed his surname in 1745, the year of the Jacobite rebellion: so perhaps we should really speak of 'Fadden's map of Norfolk', with a short 'a'. William Faden senior was a printer, best known for publishing William Dodd's sermons, books on printing and typography and pamphlets encouraging inoculation against smallpox. He was a friend of Samuel Johnson, of dictionary fame, for whom he printed his *English and Hebrew Grammar*. It is said that on his death bed Johnson asked if any of the family of Faden the printer were living. On being told that the geographer near Charing Cross was Faden's son he said, after a short pause 'as I borrowed a guinea of his father near thirty years ago, be so good as to take this, and pay it for me'.[8] William junior was apprenticed for £25 just before his fifteenth birthday to one James Wigley, an engraver in Fleet Street, and during the latter part of his apprenticeship he may (although there is no conclusive evidence) have worked briefly with Thomas Jefferys. He completed his apprenticeship, aged 22, in 1771, and was soon in partnership with Jeffreys's son, another Thomas, who was still only 18 years old and had not yet completed his apprenticeship as an engraver.[9] The young partners quickly became established in the continental trade as indicated by the survival of a series of eighty-nine letters, written in French. The correspondence vividly portrays the relationship between Faden, his partner, and a group of French, Dutch, Italian and German map sellers at the hub of the European map trade.[10] Within a year Thomas and William produced an extensive catalogue entitled *Modern and Correct Maps, Plans and Charts*. Despite inexperience and youth the pair had many factors working in their favour: they had inherited numerous map-related contacts in the Americas and Europe from Jefferys senior; and their business premises in Charing Cross (called No. 487 The Strand after the introduction of street numbering) had for many years been a major centre for those seeking to buy maps, and for engineers who wanted plans prepared for docks, canals and other civil engineering projects. Samuel Johnson described it as the busiest place in London, where the 'full tide of human existence' was to be found.[11] Most important of all, however, they inherited three outstanding surveyors: Joseph Hodskinson, Thomas Donald and John Ainslie.

The partnership was dissolved in 1776 when the younger Jefferys finished his apprenticeship and reached his majority,[12] and in the same year Faden made one of the most significant moves in his career by joining a dining club at the King's Head Tavern in Holborn called the Smeatonian Society (later the Society of Civil Engineers). This had been formed only a few years earlier by John Smeaton, the leading civil engineer of his day – builder of the Eddystone lighthouse and the Forth and Clyde Canal. The subscription was threepence a week and the publican of the King's Head was included as a member.[13] William attended meetings regularly and occasionally took the minutes. Other members included John Rennie, the bridge and canal builder; Mathew Boulton, manufacturer and engine-maker; James Watt, the inventor of the steam engine; and Major James Rennell, the surveyor of Bengal, whose maps were a model for the later Ordnance Survey. Joseph Hodskinson, one of Jefferys' surveyors, was also a member, as was the instrument-maker Jesse Ramsden, designer of the celebrated Ramsden theodolite, so vital in the precise triangulation of the country which started in the 1780s. William Mudge, later appointed as Director of the Ordnance Survey, subsequently joined; so too did Thomas Milne, the surveyor who later worked on Faden's Norfolk map.[14] These people were at the very centre of a burgeoning industrial and commercial power house, and they provided the young cartographer with an invaluable network of friends and contacts.

In June 1783, William Faden senior's will was proved. In it he left to his son 'whatever is needed to settle him in his shop', and we must surmise that a financial transaction had taken place to induce the young Jefferys to surrender the title as well as sell the business and property.[15] On the very same day Faden, now aged 34, was sworn in as Geographer in Ordinary to the King.[16] The Royal Arms could now be proudly displayed over the front of his shop (as shown on a print by Thomas Malton of Charing Cross – Figures 16 and 17) but the title bestowed few other practical advantages, simply signifying that Faden enjoyed the status of a favoured tradesman and reputable publisher, and had a sufficiently large collection of maps to fulfill the king's personal needs. While it was a coveted mark of status, it did not represent an official position in any way sanctioned and salaried by the government.[17]

Faden took up the production of county maps with vigour, almost as if he wished to prove that it was possible to make a financial success of such work, and to assuage the indignity of Jefferys' bankruptcy. He did not limit himself to England, producing many maps of Scottish and Irish counties. His involvement took two forms: acquiring and publishing maps surveyed and drawn by others; and publishing county maps *de novo*, for which he organised the surveying, drafting, engraving, printing and retailing entirely on his own account.[18] The former became Faden's particular *forte*, as he frequented sales rooms to view and bid for copperplates of a wide variety of county maps. Some of these he simply reprinted; more frequently he organised a further minor resurvey, so that plates could be re-engraved before a new print run. It has been estimated that his predecessor maintained a dozen draftsmen and engravers in his shop, and it is likely that Faden had more.[19] This method of publishing was financially less risky, for the cost of the plates was far less than that of a new survey undertaken from scratch, and his activities in this respect were remarkable. He republished Thomas Jefferys' twenty-sheet map of Yorkshire, John Rocque's eighteen-sheet survey of Berkshire, Thomas Donald's six-sheet map of Cumberland and Dury and Andrews' nine-sheet map of Hertfordshire, among others (for the full list, see Appendix 3). But equally important were the maps which Faden produced from scratch. These included Joseph Hodskinson's six-sheet 1783 map of Suffolk, John Ainslie's six-sheet 1775 map of Fife and Kinross, Thomas Milne's 1791 six-sheet survey of Hampshire and, of course, the 1797 map of Norfolk.

The American connection

In one important respect Faden's career echoed, in a remarkable way, that of Thomas Jefferys. In the case of Jefferys it was the Seven Years War (1756–1763) that gave a major boost to his business; for Faden it was the American War of Independence (1775–1783). Faden was particularly fortunate in that he inherited contacts in the map world, in both Europe and the North America, which placed him in a position to exploit the opportunities presented by the military and political situation. So great was his involvement in American mapping that he is now perhaps better known in America than he is in England. The Library of Congress in Washington has its own special Faden Collection, purchased

FIGURE 16. Thomas Malton's print of Charing Cross, 1795, with William Faden's premises shown in the centre. Reproduced with the permission of the Guildhall Library.

FIGURE 17. An enlargement of Malton's print, showing Faden's name over the shop front, complete with the Royal Coat of Arms. Opposite Faden's shop was Northumberland House, the home of Hugh, Earl Percy, who fought as a junior officer in the American War and was well known to Faden. Percy's secretary was Claude Sauthier, who was also an engineer-surveyor and produced material for some of the best American battle plans (see Figure 18).

for $1,000 in 1864, which is entirely composed of his maps.[20] The William Clements Library in the University of Michigan and the Newberry Library in Chicago, as well as a number of other public libraries, also possess extensive collections. Most of the maps they contain are related to the War, but they also include many of Faden's English county maps. With so few of the latter being in circulation they attract very high prices on the British market, for example the £2,500 asked for a copy of the 1783 Suffolk map (surveyed by Hodskinson, engraved by Faden) in 2007.

The cartographic historian J. B. Harley has remarked that the 'American revolution ranks as one of history's most carto-literate wars'.[21] There are more than 217 printed plans of battles fought during the war and it has been estimated that the British military headquarters had access to over 20,000 maps. William Faden was only 25 years old when the war broke out, and in his first year of his new business partnership: but he played a vital part in this cartographic extravaganza. His maps informed those countries directly involved in the conflict, as well as the general public in France, Britain and the Americas. A wide cross section of the British population appears to have been fascinated by the revolution in the 'American Colonies', including a sizeable minority who supported the rebels. There was thus a strong demand for maps of this unfamiliar theatre of conflict and many were published in the *Gentleman's Magazine* and the *London Magazine* alongside written reports of the action.

It is likely that Faden's maps played a part in military and political decisions made at the highest levels on both sides of the Atlantic.[22] They fell into two categories. Firstly, there were maps of large areas – colonies and provinces. These had often been surveyed some time before the war, but were now updated and re-engraved. A good example is the 1776 *Map of the Province of New York (and New Jersey Added) from actual drawings by Claude Joseph Sauthier*. Secondly, there were maps of much smaller areas directly involved in the fighting, showing troop movements and battle plans, such as his *Plan of the Operations of General Washington against the King's Troops in New Jersey 26th December 1776 to 2nd January 1777*, or the *Plan of the Town of Boston with the Intrenchments engraved by William Faden to illustrate Bunker's Hill for the British Public* or the *Battle of White Plains* (Figure 18). By 1778 Faden's catalogue listed twenty military maps covering the American campaign – and the war still had five years to run.[23]

British Officers were encouraged to visit Faden's premises to purchase maps prior to embarking for America. In 1777 he specifically produced for this purpose the *North American Atlas*, a collection of surveys of the country which was a great commercial success.[24] But like many of Faden's products this was available to the combatants on both sides, and was on sale in both London and New York. So too was his *Portable Atlas of America*, a sort of military Baedeker, or 'holster' atlas,[25] which was based on maps surveyed and published by Thomas Jefferys. We know that the staff of Rochambeau, the French commander (the French joined on the American side in 1778), were obliged to study, and purchase for themselves, the maps produced by British military engineers and published by Faden. His charts were eagerly sought after by the French *Dépôt des Cartes et Plans de la Marine*, the official repository of maps for the French Navy (the postmarks show that it took as little as three days for Faden's letters and maps to reach Paris).[26] The American rebels usually acquired

FIGURE 18. The Battle of White Plains, East New Jersey, 1776. The American forces were under the command of General Washington. Drawing by Claude Joseph Sauthier, engraved by William Faden. Reproduced with the permission of Ashley and Miles Baynton-Williams, from: *Maps of War*, London 2007.

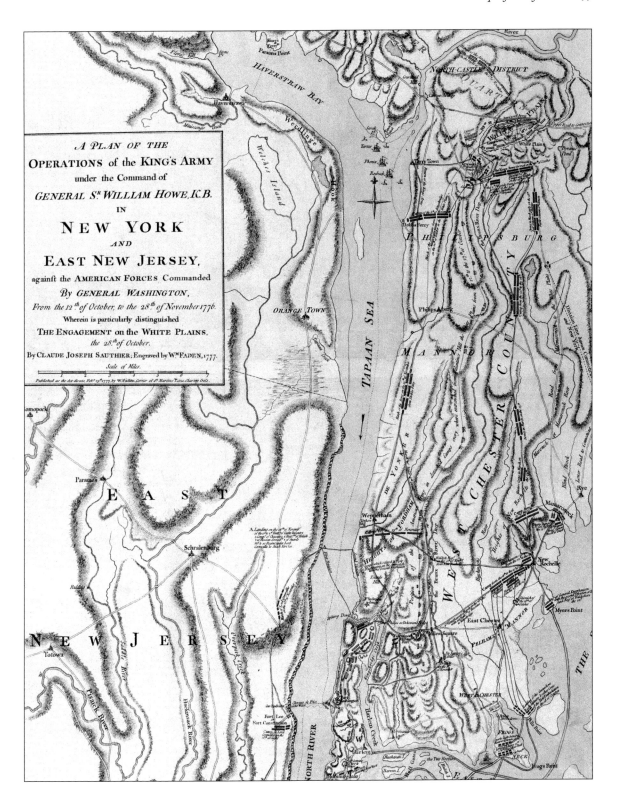

A PLAN OF THE
OPERATIONS of the KING'S ARMY
under the Command of
GENERAL Sᴿ WILLIAM HOWE, K.B.
IN
NEW YORK
AND
EAST NEW JERSEY,
againſt the AMERICAN FORCES Commanded
By GENERAL WASHINGTON,
From the 12ᵗʰ of October, to the 28ᵗʰ of November 1776.
Wherein is particularly distinguished
THE ENGAGEMENT on the WHITE PLAINS,
the 28ᵗʰ of October.
By CLAUDE JOSEPH SAUTHIER; Engraved by Wᵐ FADEN, 1777.
Scale of Miles.

their own copies of Faden's maps in Paris; this we know from the libraries and correspondence of Thomas Jefferson and George Washington, future American presidents. Washington had a particular interest in maps, having been trained as a land surveyor. In Harley's words, 'he thought cartographically … he was the most map conscious of the revolutionary commanders', and even in retirement he continued to collect maps, mainly imported from London map dealers.[27] William Faden thus made a profit, before and after the war, from the French, the British, the Americans and even from the Prussian merceneries who fought on the side of the British.

The battle plans were taken from drawings made by staff attached to, or within, the British military services. To our eyes it may seem odd that engineers and others could pass on information to private London cartographers and receive payment; but if anything the practice was actively encouraged by the military authorities. The unpredictable Atlantic crossing could be made, in the right conditions, in no more than four weeks and some of Faden's maps of military actions were published within six weeks of their occurrence. The *Attack and Defeat of the American Fleet under Benedict Arnold by the King's Fleet Commanded by Sir Guy Carleton upon Lake Champlain on the 11th October 1776*, 'from a sketch taken by an officer on the spot', was thus published in London on 3 December. Even more extraordinary is the way in which Faden's contacts in Paris purchased his American maps from London, translated the English captions to French, changed the cartouche wording of 'American rebel' to 'American soldier', added hydrographic details that could only have come from the French Admiralty – and then offered them for sale to Faden himself![28] The process worked the other way as well. One French map-seller wrote to Faden: 'please do not send maps that you have copied from the French ones'. The map-sellers on both sides of the conflict knew exactly what was going on, and relations cooled slightly only when France joined the war in 1778. The cross-channel trade in maps still continued, but now had to go via the free port of Ostend rather than through Calais. Some orders were huge – in one case for more than 1,200 maps at about £1 each – but many were for single copies, maps which may themselves then have been copied.[29] There is no evidence that sensitive material was passed on to France by Faden, or that secret French admiralty information came Faden's way. Faden was of course Geographer to the King and must have been keen to preserve the prestige of this official position. Certainly, he would have been aware of the potential risks: he was called as witness in the notorious trial of Francois Henri de la Motte, who was executed for high treason in July 1781. Living in London, but previously a French army officer, de la Motte was found guilty of passing information about British fleet dispositions at Portsmouth and Port Praya Road, St Iago (Cape Verde Islands), an action that indirectly led to the death of 207 British sailors. Faden was called by the defence and admitted to selling maps and admiralty charts to de la Motte in 1779 and 1780 at a cost of over £90. It transpired in court that these maps were readily available to the public and that 'they were not a secret to the Lords of the Admiralty'.[30] The judge passed a chilling sentence '*pour decourager les autres*':

> That he should be hanged by the neck, but not till he be dead, then to be cut down, and his bowels to be taken out and burned before his face; his head to be taken off, his body cut into four quarters and to be put at his Majesty's disposal.

Only a few months before the end of the American war 80,000 people attended his execution at Tyburn; the story was subsequently adapted by Charles Dickens in *A Tale of Two Cities* and in Thackeray's last, unfinished, novel, *Denis Duval*.

As one might expect, the large collection of William Faden's maps of American battles in the British Library (Appendix 4) tends to concentrate on British victories, real or perceived, whereas in the collection (larger, with 101 items) in the Washington Library of Congress American successes tend to feature most prominently. With his semi-official position Faden was of course keen to report the good news to the British public. But it is noteworthy that, ever the astute businessman, towards the end of the war (and in anticipation of future markets) he changed the word 'rebel' to 'American' on his copperplates.

Not all of the maps published during the war were of equal value to those people involved in the conflict. Some were based on surveys taken twenty years earlier, sometimes using different meridians as their base, with added confusion over true and magnetic north. Indeed, Thomas Gage, the British commander in chief, went so far as to assert that 'there is not a map of the inhabited provinces of any use, there is none correct, even the roads are not marked'.[31] Thomas Jefferson, on the other side of the conflict, described a map of Virginia in 1780 as 'a mere cento of blunders, it gives only a general idea of the courses of rivers and the positions of counties'.[32] With the end of the war this phase of Faden's cartographic career came to a close, although not completely, as his American topographic maps remained popular, and as a sideline he continued to publish maps to accompany written histories of the conflict often penned by military commanders keen to revive tarnished reputations, including Generals Burgoyne (defeated at Saratoga) and Cornwallis (defeated at Yorktown). His American contacts continued when Thomas Jefferson, while Minister for France in Paris, paid a visit (on 17 March 1786) to William Faden's establishment in Charing Cross. He purchased some American maps but the main reason for his visit was to discuss a map he had himself drafted to accompany his book *Notes on the State of Virginia*. He asked Faden how much he would charge to engrave his map and the reply was 'fifty guineas'. He then walked on a few yards to No. 352 The Strand, to another engraver, Samuel Neele, who offered to do it for 'between twenty and twenty five pounds'.[33] Understandably he chose Mr Neele, but he came to regret his decision; Neele's engraving had 172 errors and he charged £28 16s 9d! Jefferson was so displeased he took the plate to Paris for re-engraving; the French invoice states that this cost a further 110 livres as a further twenty-five days' work were required. The story does not end there. Jefferson wanted his map to be published in London by John Stockdale and the latter suggested that William Faden, 'a tradesman of the strictest honour and integrity in his line of business', was the man to do it. Faden would be allowed to print and sell the map, provided that he paid a fee per copy. He offered 7d per map but Jefferson wanted 10d and so the work went to Cox's, printers in Chancery Lane.

Faden's sister Hannah ('Henny') married the painter John Russell after some resistance from William Faden senior, who did not approve of his strong nonconformist leanings.[34] Russell, also an astronomer and amateur globe-maker, became an eminent portrait painter and his picture of the younger Faden (reproduced on the front cover) shows the face of a 'man of extraordinary energy, technical ability, business acumen and sheer determination'.[35] The portrait now hangs in the Map Library of the British Library. Even the American Governor, Thomas Pownall, spoke of him as a 'very industrious and accurate young man, skilful in engraving'.[36]

Despite all his involvement in the events across the Atlantic, Faden continued to publish maps of English counties. He bought up the plates of John Chapman's four-sheet Nottinghamshire map for forty-five guineas in 1784, and of William Yates's six-sheet Staffordshire map in 1789. He was asked to value, for probate purposes, the copperplates left from Lindley's map of Surrey: he offered only £50 'in their present worn out state', yet he was still profitably printing from them twenty years later (indeed, the same plate was in continuous use until 1874, forty years after Faden's own death). He was accused, while doing the valuation, of surreptitiously taking an impression. He was incensed by this accusation, writing: 'you will please call for the settlement of your account and the sooner it is discharged the better it will be'.[37] Little by little, he came to dominate the trade in large-scale county maps; Paul Laxton has calculated that of the forty-three examples published between 1748 and 1800, Faden acquired the engraving plates of no less than thirty-one. Together with those new surveys which he himself commissioned, Faden owned the sole rights to three-quarters of the series published by the turn of the century.[38]

Nor were his publications, by the later stages of his career, limited to maps of America and the English counties. He produced surveys of Portugal, Africa and India and he is even credited by some as being the first to use the name 'Australia' on one of his maps, rather than the usual New Holland or Hollandia Nova. His maps were directly copied to produce the first Muslim-published atlases. The *Cedid Atlas Turcusemi* (the New Atlas), in which Faden's maps were overlaid with Arabic script, was published in Istanbul.[39] In 1807 he published the second edition of Richard Horwood's extraordinary 32-sheet map of London, first engraved and published in 1799. It was surveyed at twenty-six inches to the mile and was so accurate that Hackney carriage fares were calculated from it (Figure 19). He continued, throughout his career, to maintain a close relationship with the Society of Arts. In 1794 it was discovered that several county maps were missing from their collection, and he promised to supply them at no cost. In total he donated copies of eleven surveys, and offered to

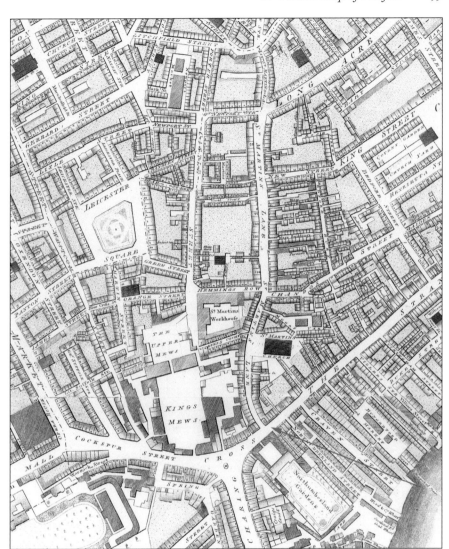

FIGURE 19. Part of Richard Horwood's 1799 Map of London, drawn at 26 inches to the mile and later published by Faden. Faden's shop at 487 The Strand is shown by the red arrow. Reproduced with the permission of Patrick Mannix of Motco Enterprises: www.motco.com.

supply some others that were out of print.[40] One suspects that his failure to gain a prize for his Norfolk map was not too great a setback to his flourishing career. His connections with the Board of Ordnance were close and he published, at his own expense, successive volumes of Mudge and Dalby's *An Account of the Operations carried out for accomplishing a Trigonometrical Survey of England and Wales 1799–1811*.[41] He produced the first published Board of Ordnance map, *An Entirely New and Accurate Survey of the County of Kent*, on 1 January 1801, exactly ten years after the Trigonometrical Survey had been formally established.

At about this date Faden moved his business to larger premises just a few doors away at No. 5 Charing Cross. He retired to Shepperton in 1823, the business passing to James Wyld, who had served as his apprentice from 1804 to 1811, and who in turn became Geographer to His Majesty and his

Royal Highness the Duke of York. James's son, also a James, was later made Geographer to Queen Victoria and Prince Albert.[42] A catalogue made at the time of Faden's retirement lists no less than 350 maps; and so successful had he been in acquiring county maps that, by this stage, he had in his personal possession plates for no less than twenty-eight of them.[43] He died in 1836 a wealthy man, leaving £20,000 – which we may compare with the £20 left by his predecessor Thomas Jefferys.[44] He was buried by his nephew, William Russell, the Vicar of Shepperton; his memorial can still be found in the parish church there. It seems that he became a little confused in his latter years and wrote many codicils to his will, leaving bequests to the Middlesex Hospital (founded in Windmill Street, off Tottenham Court Road, in 1745, four years before Faden's birth), to the Hospital for Fistulas and to various charities for orphans. These codicils were challenged in the years after his death by his executors, including his nephew William, who claimed that he had been of unsound mind at the time.[45] Faden's premises in Charing Cross were demolished in 1830–2 to make way for the development of Trafalgar Square.[46]

The surveyors: Thomas Donald and Thomas Milne

Faden's map of Norfolk was surveyed by two men of exceptional experience and ability who almost certainly already knew each other. Thomas Donald originally came from Anthorn in Cumberland and worked mainly as an estate and town surveyor, although he is best remembered for his county maps. He was involved with John Ainslie in a survey of Buckinghamshire, made between 1766 and 1768, which was published by Andrew Dury and later by Robert Sayer. Donald, who worked as Ainslie's assistant, was only 15 or 16 years old at the time: Ainslie himself was only 21. He worked with Ainslie again, and with Joseph Hodskinson, in 1767–70, on a survey of Yorkshire, which was published under Thomas Jefferys' name but after his death; and again with Ainslie between 1770 and 1771 on a survey of Cumberland, engraved by Joseph Hodskinson and published by Thomas Jefferys and later by William Faden (this map is usually referred to as *Donald's Map of Cumberland*). The following year the two men again collaborated, this time on a survey of Westmorland. Donald also surveyed the neighbouring parts of Yorkshire and, with the increasing importance of the tourist trade in this part of the country, subsequently produced a *Map of the Environs of Keswick* in 1789. He worked with Ainslie yet again in the 1790s, producing a survey of Bedfordshire which was engraved by Joseph Hodskinson and first published by Thomas Jefferys in 1795. In total, Donald worked with Thomas Jefferys for seven years and with John Ainslie for six. Before he undertook the Norfolk survey he had also worked with Lieutenant Murdoch MacKenzie junior, maritime surveyor at the Admiralty, on a coastal survey of the Bristol Channel, south-west England and the Thames estuary, which was carried out during the 1770s and 80s.[47]

Thomas Milne's early career was spent surveying the 'forfeiture' estates in Scotland – the properties of Jacobite supporters which had been confiscated after the 1745 rebellion, the thirteen largest of which were annexed by the Crown. Prior to 1784 he worked, as assistant to Peter May, for the Earl of Fife and the Duke of Gordon, and on the Strathbogie estates. As this Scottish work began to dry up he moved south, and worked from 1788 to 1790 as surveyor on the county map of Hampshire, published by Faden in 1791, the quality of which was duly recognised by a prize from the Society of Arts. The map is noteworthy for the care taken in the depiction of the process of enclosure, the labelling of open fields, the differentiation made between open and enclosed roads and the distinction carefully drawn between downland and heath.[48] It records the names of more than 250 landowners but is less successful in its portrayal of small-scale industry. By the time that Milne worked on the Norfolk map, from 1790 to 1794, he was a member – like Faden himself – of the Smeatonian Society.

Milne's greatest contribution to cartography in fact came after his work in Norfolk had ended. He applied for a post with the Ordnance Survey but for unknown reasons was turned down. Putting this disappointment aside, he began work on a comprehensive map of the capital, *The London Land Utilisation Map*, or more fully, *Milne's Plan of the Cities of London and Westminster, circumadjacent Towns and Parishes*, which he surveyed, engraved and published on his own in 1800 (Figure 20).[49] Surveying began in 1795, the year after his work in Norfolk had been completed, and in pictorial terms the map draws strongly on Milne's

FIGURE 20. An extract from Thomas Milne's London Land Utilisation Map of 1800. Reproduced with the permission of the British Library.

background in estate surveys. Drawn at two inches to the mile and engraved on six plates, it covers 260 square miles of the London area and shows, using colours and hatching, seventeen different types of land use, including arable, meadow, pasture, market gardens, woods, marshes, open fields and commons (one-third of London's agricultural land still remained unenclosed at this time).[50] The map allows us to see how Londoners were supplied with their meat, grain, fruit and building materials, as well as where they could seek their leisure. The landscape it shows is now almost unimaginable; alluvial meadowland near Deptford and Dulwich, market gardens in Bow, Hampton, Isleworth and Brentford, osier beds for basket-making on the river banks between Fulham and Staines and barley and wheat fields in Hammersmith and Chelsea. As Foxell has remarked, 'Milne's breakthrough was the graphical description, through the use of colour-coding, of differing land use and intensities, as well as the age of settlements – he was the first to map social data accurately in a way that abstracts it from the purely geographical.'[51] For modern students, the map has the added advantage of indicating the kinds of soil which underlie London: soil maps of urban areas are almost impossible to produce (the land surface being so completely obscured by roads and buildings), but the kinds of land use mapped by Milne provide a useful surrogate. It was 130 years before a similar project was attempted, in L. Dudley Stamp's Land Utilisation Survey of the 1930s.[52]

Milne's map of London was the first to be published using official triangulation data, following the publication of General William Roy's precise base-line measurements on Hounslow Heath in 1790. His map overlaps with that of Kent which William Faden prepared for the Ordnance Survey: immediately after their work together in Norfolk had ended, in other words, Faden and Milne were working in the same part of the country but on different projects. For reasons which remain obscure Milne's map exists only as a single copy, in the King George III Collection in the British Library. Each sheet is signed by Milne himself, but has a serial number, so it is just possible that other copies exist. The map's extreme rarity could be put down to the Board of Ordnance objecting to its publication on copyright grounds, but this would have been unusual, as soon afterwards that organisation actively encouraged county surveyors to employ their triangulation data. A more likely explanation is that objections were raised by Richard Horwood, the surveyor of the 26 inches to the mile 1799 plan of London, which was also published by Faden. Many of the features on Milne's map are almost identical to those on Horwood's, and the latter may well have felt aggrieved.

Milne's status as a cartographer and surveyor is indicated by the fact that he was asked to write two chapters in George Adams's *Geometrical and Graphical Essays*, the standard text used by all surveyors towards the end of the eighteenth century.[543] These were entitled 'Mr Milne's method of surveying with the best theodolite' and 'Mr Milne's Observations on Plotting'. Adams (who had been a supporter of and subscriber to Benjamin Donn's map of Devon) described Milne as 'one of the most able and expert surveyors of the present day … his

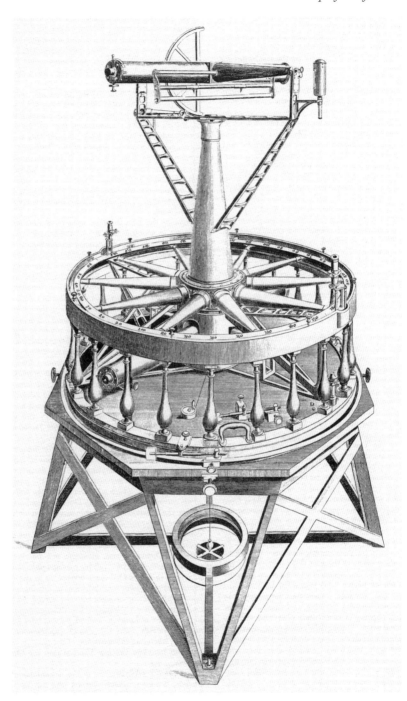

FIGURE 21. Jesse Ramsden's 'Great Theodolite', completed in 1787 and drawn here by Thomas Milne. The illustration first appeared in the Philosophical Transactions for 1790.

method deviating from the common practice'. Milne also drew and engraved the plates in the volume showing surveying instruments and specimen surveys of estates (Figure 21).

No records survive to tell us which parts of Norfolk were surveyed by the two principal surveyors, nor can we be sure whether they were working in the county at the same time (in the case of Jefferys' 1771 map of Yorkshire three surveyors were employed and we know from his surviving papers who was responsible for each area).

Captain Joseph Huddart

At the bottom of Faden's map of Norfolk is the note: 'The Nautical Observations in Yarmouth Roads and Happisburgh Gatt are laid down from a survey taken by Capt. Joseph Huddart by the order of the Trinity House, Deptford Strond.'[54] Huddart (1741–1816) was a man who typifies the atmosphere in which Faden and his colleagues lived; intellectual, dynamic but essentially practical. Born in Allonby in Cumberland, Joesph Huddart went to sea in his teens, sailing frequently to North America, and eventually took command of a brand new vessel, the *Royal Admiral*, for the East India Company, travelling extensively to China, India and Sumatra via the Cape of Good Hope.[55] Interested in astronomy and mathematics from a very early age, he wrote a series of letters to Nevil Maskelyne, the Astronomer Royal, on a method of fixing longitude off the ports of Bombay and Calcutta using observations of the eclipses of Jupiter's satellites. On retiring from active service with the Company he became involved in hydrography, inventing the three-arm protractor and charting the west coast of Scotland and the St George's Channel (between Rosslare in Ireland and St. David's Head in Wales). In 1790 he was asked by Trinity House (the body responsible for lighthouses and coastal navigation) to carry out a survey of the shoals off 'Hasborough Gatt and the Newarp Bank' in east Norfolk, with a view to establishing a floating lighthouse there to act as a warning for shipping (Figure 22). The lighthouse had been commissioned by the time Faden's map was published, in 1797, and is noted on the map as lying a 'distance of two leagues from, and in line with, Martham church and Winterton Great Light'. Huddart's charts and the lighthouse were unfortunately not able to prevent one of England's worst naval disasters. On 16th March 1801 HMS *Invincible*, a ship of seventy-four guns, on its way to support Admiral Nelson in the attack on Copenhagen, became grounded on Hammonds Knoll, a sandbank lying east of the Happisburgh Sands. *The Times* reported how 'she beat most violently for more than two hours … the mizzen mast went by the board and the main mast was immediately cut away'. More than 400 of its crew of 595 were drowned and many of the dead were buried in a mass grave in Happisburgh churchyard.[56]

Huddart was one of the foremost hydrographers of his day and his charts were commissioned, engraved and sold by Robert Sayer. He was yet another member of the Smeatonian Society and became actively involved in designing lighthouses (including South Stack off Anglesey), harbour improvements (Hull and Swansea) and the East India Docks in London, for which he laid the foundation stone when William Pitt, the Prime Minister, was unable to attend the ceremony. In 1777 he wrote a paper on colour blindness which had to be read to the Royal

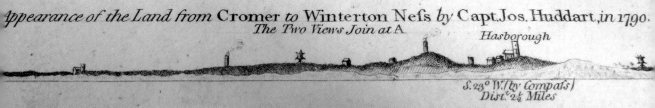

ppearance of the Land from Cromer *to* Winterton Nefs *by* Capt. Jos. Huddart, *in* 1790.
The Two Views Join at A
Hasborough
S. 23° W. (by Compals)
Dist.ᵉ 2¼ Miles

FIGURE 22. Joseph Huddart's chart of the Norfolk Coast, surveyed in 1790. From *A New Chart of the East Coast of England* published 12 May 1794 by Laurie and Whittle, Fleet Street. Reproduced courtesy of the National Archives.

Society by the Rev. Joseph Priestley, as he had gone back to sea. He showed how the condition was familial, and he was the first to use coloured objects to test for its existence.[57] Fifteen years later he read another paper to the Royal Society on the phenomenon later called mirages – the visual disturbance caused by the refraction of light on the horizon under certain atmospheric conditions.[58] He had experienced the effect himself while sailing off Macao on the coast of China. Huddart is, however, best remembered for the very detailed research he carried out on cordage, or heavy ships' ropes. He noted that the outer fibres of the anchor cable broke first when rope was severely stretched. He designed and manufactured, initially in his own garden, a machine that distributed the strain to all layers of the yarn cord, thus increasing the rope's tensile strength. He patented the machine, went into production and eventually became extremely wealthy.[59] His name is perpetuated in those of two streets, Huddart Street and Joseph Street, in Limehouse, East London. His manufactory in Limehouse is clearly shown on Richard Horwood's 1799 map of London, engraved by William Faden.

Huddart had another direct connection with Faden's map of Norfolk, this time on the western side of the county. The River Ouse made a long and circuitous journey for five miles above King's Lynn to Wiggenhall St German, and the adjacent land was frequently flooded and difficult to cultivate. On some occasions, corn had to be reaped and fruit gathered from platforms moved along by boats, and the ground was so soft that horses were fitted with wooden boards attached to their feet to prevent them from sinking into the soil.[60] In addition, shifting sandbanks in the river itself were a hindrance and hazard to shipping. In 1781 parliament passed the Eau Brink Act, which sanctioned the construction of a new watercourse which would bypass this stretch of the river. Two schemes were proposed, so radically different that Huddart was asked to adjudicate between them (Figure 23).[61] At this point another county surveyor, Joseph Hodskinson (of the 1783 Suffolk map), became involved. He was asked by a parliamentary commission to represent the Mayor and Corporation of King's Lynn, who were concerned that the construction of the new cut would lead to increased silting of Lynn harbour. He made his report to the House of Commons in 1794 and the Bill for the scheme was passed in 1795. By the time that Huddart was consulted the litigation fees had already reached £80,000. Huddart's suggestions were accepted and John Rennie was appointed engineer-in-chief. The new channel, which reduced the navigation distance by two miles and the water level at St German's bridge by five feet, was not completed until

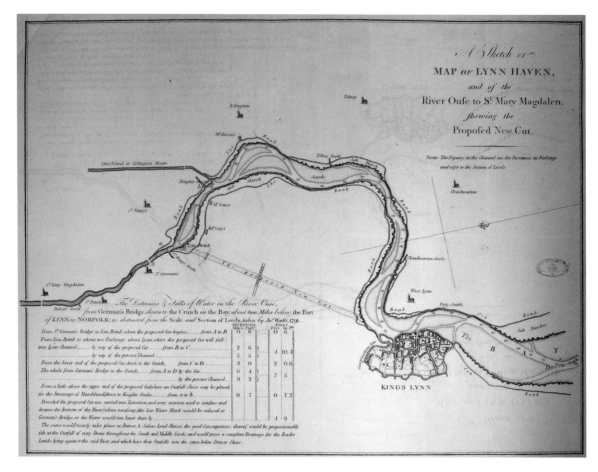

FIGURE 23. Plan by John Watté of the proposed New Cut on the Eau Brink. Reproduced courtesy of the National Archives.

1821.[62] The proposed cut is nevertheless shown on Faden's map and the two versions in the National Archives have been re-engraved, with details of the work proposed added to the right of the map of King's Lynn.

When he finally retired, Huddard set up an observatory in his home in Islington, London. He designed and constructed, entirely on his own and using a forge in his garden workshop, a five-foot equatorial telescope. On his death he left no less than £60,000, together with 7,000 acres in the Porthmadog district of North Wales, an area he had passed through on his way to design the South Stack lighthouse.[63]

Mostyn Armstrong and the map that never was

On 30 November 1776 an advert appeared in the *Norwich Mercury* announcing proposals for a new map of the county to be surveyed by Captain Armstrong and his son.[64] Andrew Armstrong (1700–1794) was a surveyor with an army background (in 1768 he is described as a Lieutenant on half pay from the 32nd Regiment)[65] who had already worked on a number of county maps. Helped by his son, Mostyn John, he surveyed Northumberland between 1766 and 1769, producing a map for which he was awarded a Society of Arts prize of £50. The two men also worked together on surveys of Berwickwickshire and the Lothians, in 1771 and 1773 respectively, while Mostyn on his own surveyed and published Peebleshire in two sheets in 1775, and subsequently produced two road books – *An Actual Survey of Great Post-Roads London to Edinburgh* (1776) and *An Actual Survey of Great Post-Roads London to Dover* (1777), which went through several editions – together with a *Scotch Atlas* of thirty maps in 1777.[66] The topographer Richard Gough had a poor opinion of this latter work, commenting that:

> the Atlas is little valued … his pretension to actual survey is entirely chimerical: he copies others, in grafting mistakes of his own, and run over the counties in a strange cursory manner … Armstrong has attended to his own, and the engraver's profit more than that of the public or their information.[67]

Andrew, meanwhile, worked alone on a survey of the County Palatinate of Durham, published in four sheets by Thomas Jefferys in 1768; on a map of Ayrshire in six sheets, published in 1774; on a map of Lincolnshire in eight sheets, published in 1779,[68] which sold just under 500 copies; and on one of Rutland, published in 1781. Both Andrew and Mostyn were members of the Society of Arts and originally lived in the Scottish borders before moving together to Norwich in 1775 or 1776.

Detailed proposals for their Norfolk map appeared in the *Norwich Mercury* between May and July 1777, but after this the father's name is no longer mentioned. Further reports in the Norwich newspapers stated that surveying had commenced in May 1777, and that work was to be completed by August 1778, to be followed by publication in London in December of the same year. Mostyn Armstrong estimated the total cost of the survey at £450, clearly an underestimate, and that the price of the map would be two guineas to subscribers – one guinea on subscription and one on delivery. The number of subscriptions already numbered 50 and would, he said, close at 300, at which point the price per map would be increased to £2 10s.[69] In November 1777 Armstrong reported that he had completed the surveys for the two western sheets (there were to be eight sheets in all) and that the drawings of these were being prepared for the engraver. Proof copies of one sheet ('coast and county from Hunstanton lighthouse and to Lynn Regis, Wisbech and Pentney Abbey') were made available for inspection from April 1778 at booksellers and coffee houses in Norwich and Lynn, at which point the subscription list stood at 220. Towards the end of December 1778, by which time there were 320 orders for the map, a sketch appeared in the *Mercury* showing the proposed layout of the published survey – the county was to be mapped in two rows of four sheets each. Of these, 'sheets 1 and 5 [the west of the county] are engraved, 2 is nearly ready for it, 6 is in great forwardness and a considerable proportion of the other four sheets are done' (see diagram below).[70]

1	2	3	4
5	6	7	8

At the same time Armstrong, then living at 2 Redwell Street, had taken on the work of editing and publishing a multi-volume historical work, *A History and Antiquities of the County of Norfolk*, which was to be a corrected, abridged and enlarged version of Blomefield's *History of Norfolk*, previously published between 1736 and 1752. Mostyn's book was eventually printed as ten volumes, the first of which was published in 1781. The historian Walter Rye later commented that 'of the impudent plagiarism of Blomefield's History by Mostyn Armstrong it is not worthwhile to speak'.[71]

Meanwhile, the new county map was making slow progress. Mostyn seems to have become distracted from his task, in part perhaps because he had been appointed a lieutenant in the Norfolk militia by Lord Orford, to whom the map was to be dedicated.[72] He also became involved in organising a group of Norwich intellectuals called the Society of United Friars, who read scientific and philosophical papers to each other and arranged lectures on subjects as diverse as the prevention of bladder stones, experiments on dyes, ventriloquism and the theory and practice of electricity. Their unusual name stems from the practice of wearing the habits of a monastic order. Mostyn was Abbot (chairman) in 1786, followed by a period as Procurator (secretary and recorder of minutes). The legibility of his minutes deteriorates markedly during 1787; there are multiple corrections and crossings-out, and splashes of ink cover the pages. This may suggest a period of illness but in the year before his death his writing had returned to normal.[73]

During this period there were mutterings of discontent in the Norfolk papers and questions were raised about the £300 that had already been taken in subscriptions. A committee was eventually set up in 1786 to enquire into the affair and on 19 May 1787 it announced that 'in consequence of the engagement with Mr. Armstrong being dissolved the Committee had accepted the offer of Mr. Faden to carry out a new survey', which was to be printed in six sheets at a cost of two guineas to subscribers.[74] The Norfolk gentry were not going to be hoodwinked again, however: they insisted that Faden – famous London cartographer though he might be – would have to present completed survey drawings before any money changed hands. Armstrong replied the following week, describing William Faden as 'a retailer of Geography from London' and reiterated 'his determination to pursue the work to its final conclusion'. But in October 1789 the *Norwich Mercury* announced the sale of the contents of Armstrong's house at St Martin at Palace, including maps, plans, prints and a considerable library.[75] From what we know of William Faden it is inconceivable that he did not know of the sale, and while we have no direct evidence that he purchased the relevant survey drawings or copperplates (if such indeed existed) it is quite likely that he did so. Armstrong died on 11 December 1791, by which time Faden's surveyors had been working in the county for almost two years.[76] Armstrong's is a sad story; he also published proposals in 1778 to map the county of Cambridgeshire, a project which likewise came to nothing. Had he completed his Norfolk map it would have been a reasonable piece of work, but it is unlikely to have equalled that produced by Donald and Milne.

One piece of his work does, however, appear to have survived. Henry Swinden made a survey of Great Yarmouth in 1753 and on his death in 1772 this passed to John Ives junior, his friend and patron. Ives was a member of Mostyn's wife's family and he allowed Armstrong to make a copy of the survey in June 1779. This was published in August 1779 by 'Mostyn John Armstrong – County Surveyor' at the price of 3s. 6d.[77] When this survey is compared with the drawing of Great Yarmouth included (at the bottom right corner) on the Faden map it is clear that the two are one and the same.

Urgent military demands

With the outbreak of war with France in 1793 the acquisition of adequate maps of the country became a matter of great urgency for the government. Accurate surveys of the whole of the south coast, and of Essex and Suffolk – the main areas at risk from invasion – were already available to the Board of Ordnance and the Quartermaster General. Indeed, the officials at the Drawing Room at the Tower of London had made it their business throughout the 1790s to purchase, from William Faden, the relevant maps.[78] But there was as yet no decent map of Norfolk, and the military surveyors who were available to the Board of Ordnance were simply not up to the task of plugging this gap with sufficient speed. In 1795 the Quartermaster General appointed one Captain Thomas Reynolds, 'at an allowance of 20 shillings a day to be paid by the Ordnance, his commission signed by His Majesty', to obtain such a map.[79] At this point in time the drawings for William Faden's map, which had been surveyed between 1790 and 1794, lay in his office in Charing Cross, awaiting the engravers. The Quartermaster General authorised Capt. Reynolds, quite independently of either the Board's Trigonometrical Survey or the Interior Survey:

> To engage Mr. Faden, of the Strand, to furnish government with a correct copy of his survey of Norfolk, for which he is to be paid one hundred guineas when the same is delivered. It may be necessary to observe that there is no map published that gives any material information of the county of Norfolk, that the engraving of Mr. Faden's survey cannot be ready for delivery before next year, and that in the meantime the copy he is to put into the hands of government within the space of three weeks will contain much of the Detail necessary for Military Operations[80]

Faden acceded to the request but, mindful as ever of his commercial interests, replied:

> Mr Faden is willing to give the government a copy of his original Map of Norfolk … conditionally that it is not made public previous to the Engrav'd copy coming out, as he has already sunk £2000 in carrying on the survey[81]

This map, entitled *Norfolk – Map of the coast between King's Lynn and Great Yarmouth, and areas up to about 12 miles inland at 1 inch to 1 mile 1793*, survives in the National Archives (Figure 24).[82] That it was indeed based on the survey work recently carried out by Donald and Milne is clear from Figure 25, which

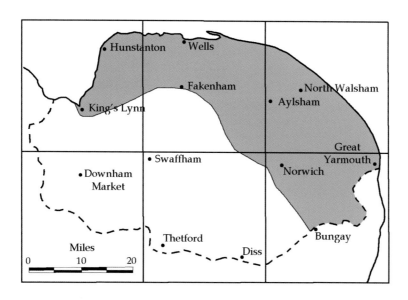

FIGURE 24. The section of the Norfolk coast shown on the 1795 military survey is shown in red.

FIGURE 25. The section of the Norfolk coast for which a map was requested by the military authorities in 1795. Reproduced courtesy of the National Archives.

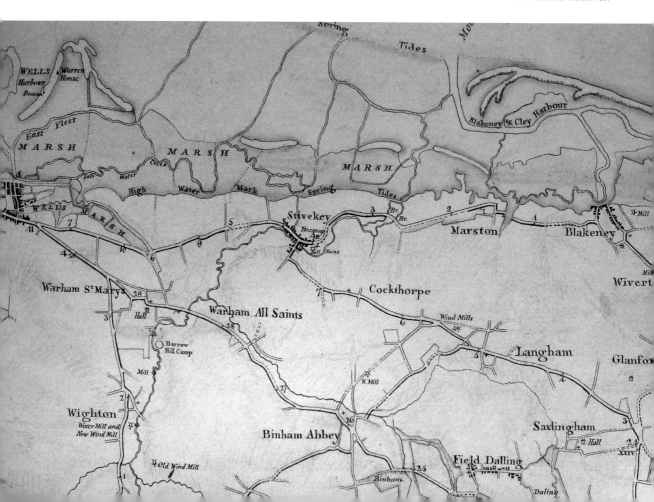

shows its treatment of the area around Stiffkey and Blakeney on the north coast, compared with the depiction of the same area on Faden's published map (Figure 26). The coast is shown in great detail but, inland, roads, parks, commons and woods are depicted in outline only, while the hachures and rivers are in light pencil. The spelling of place names is erratic, including such errors as Stivekey, Marston, Oxned and Scottaw. The date of 1793 does not exactly fit with the Ordnance Board's letter of 1795; it presumably reflects the date at which the survey was carried out, rather than that of the engraving. It is unclear whether the map was hurriedly prepared, 'within the space of three weeks', by Faden's draftsmen and engravers at the demand of Capt. Reynolds, or whether Faden, with his ear ever to the ground, had the map drawn months, if not a full year, in advance, in anticipation of such a request from the government; but the map certainly does not have the appearance of a hastily prepared work. It is perhaps unique among all county map ephemera; it is generally stated that no field drawings or drafts have survived from the whole county map series, but this would seem to be the one exception.

FIGURE 26. The same area of coast (as in Figure 25) as it appears on Faden's map of 1797.

The fee of 100 guineas received by Faden was, by coincidence, the same as the prize offered by the Society of Arts for county maps. Faden's co-operation with the Board of Ordnance was noted and within a few years (in 1801) he was asked to engrave and sell the first Ordnance Survey map, that of Kent.

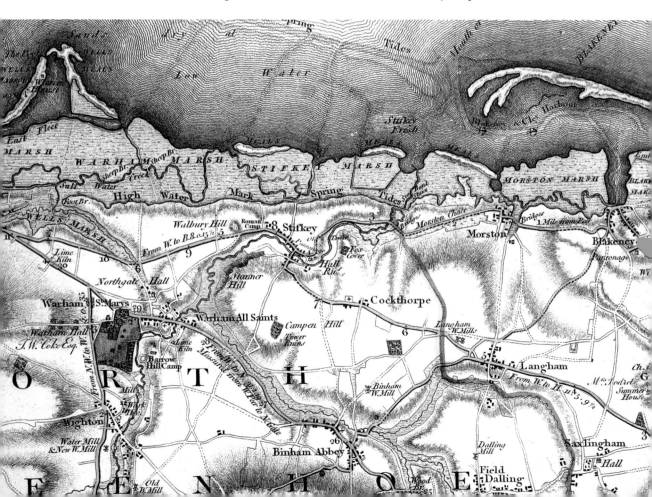

Publication

No records have survived of the progress of Milne and Donald's survey work in Norfolk, but this should occasion no surprise; as already noted, no field drawings, notes or diaries relating to any of the county surveys have yet been found. But survey work had been completed by 1794, after which the map was then a further three years passing through the hands of draftsmen, engravers and printers in Faden's workshops before final publication in 1797. It measured 66½ by 48 inches and was issued either in six sheets or – more usually – dissected, mounted on cloth and folded. It bore no elaborate cartouches, drawings of agricultural scenes or heraldic symbols. Faden included Welsh feathers in the compass drawing, perhaps to indicate his position as Geographer to the Prince of Wales, but the map is otherwise rather utilitarian in appearance.

There must have been a collective sigh of relief on the part of all concerned when the advert announcing the map's arrival appeared in the *Norfolk Chronicle* on 19 August 1797:

> A New Topographical Map of the
> COUNTY OF NORFOLK
> surveyed and measured in the years 1790, 91, 92, 93 and 94.
> by Tho. Donald, Tho. Milne
> AND ASSISTANTS.
> Planned from a scale of one inch to a statute mile.
> Executed and published at the expense of the proprietor,
> William Faden, Geographer to his Majesty, and to H.R.H.
> the Prince of Wales.

> In the above map are described all the seats of the nobility and gentry, woods, parks, heaths, commons, rivers, great and cross roads, marsh and fen lands, market towns, parishes, villages, farms etc. Also the remains of Roman roads, camps and other antiquities; embellished with plans of the towns of Great Yarmouth, Lynn and Swaffham. The map is printed on six sheets of the largest atlas paper. Price to subscribers is Two Guineas and a half, in sheets; to non-subscribers it will be three guineas.

> Subscriptions are received in the county by Roger Kerrison and Bartlett Gurney Esq., at their several Banks and by the proprietor of the map, W. Faden, Charing Cross, London. The subscription will close on the 31st day of December next, when a list of subscribers will be printed. *Note* subscribers' maps will be ready for delivery on the 1st day of October next; and to non-subscribers on the 1st day of January 1798.

> Charing Cross August 5 1797.

Several of the county maps came with an index map – a smaller version of the whole – and sometimes with a list of subscribers. Neither has been found for the Norfolk map. Moreover, compared to other county maps, this map, perhaps because it was published comparatively late in the sequence, was not extensively revised and republished. This contrasts strongly with maps such as Yates's survey of Staffordshire, which Faden reissued in 1799 twenty-five years after its first appearance, with extensive re-engraving of two plates and minor changes to others. Faden's Norfolk was in fact published in only five main editions.

Firstly, there is the 1797 original, at a scale of one inch to the mile, of which a number of versions, with minor differences and additions, are known. The National Archives thus have two, slightly different from each other, which include an 'Extract from the Eau Brink Act of Parliament' just to the right of the town map of Lynn;[83] while one of the present writers (ADM) has in his possession a version which differs from the main publication only in that Shotesham Park, to the south of Norwich, has been extended slightly to the north. Secondly, there is Donald and Milne's one inch to two miles edition, published on 12 August 1803. In this smaller-scale version much of the relief has been omitted, as have names of owners and the inset maps around the margins. This was reissued in 1825 by J. Wyld senior, and again in 1840 by his son, also J. Wyld. The final version was published in 1872 with parliamentary divisions and railways added.

None of the copperplates for the original map have survived: indeed, none survive for any of the eighteenth-century county maps, with the exception of a single six-inch-square fragment of the north-east quarter of William Yates's 1793 map of Warwickshire, which was found in a garden at Claverdon in 1955.[84]

Why no prize?

With his past record of county map awards and his close relationship with the Society of Arts, Faden must have been fairly confident when he wrote to the Secretary of the society on 31 March 1798:

> I have herewith taken the liberty of troubling you with a new topographical Map of the County of Norfolk … the only map of the County extant which has been constructed from actual measurement … I request Sir, you will have the kindness to lay it before the Committee … for their examination and patronage.[85]

Accompanying his letter were two from gentlemen supporters, Thomas William Coke of Holkham[86] and Joseph Windham of Earsham Hall. The latter described the map as 'the best executed I have seen … I have no doubt Mr Faden will receive the credit for it he is justly entitled to'.[87]

The Committee, however, were not happy, and on 7 April requested that:

> Mr Faden communicate to the Committee the information by which he claims any and to what degree superiority in the map; and that he also be requested to obtain any such information from the gentleman who has surveyed the County under the direction of the House of Commons, and also from the Post Master General as may supply the Committee with corroborating evidence in favour of the improvement by this map.[88]

Three weeks later they wrote again:

> The survey is very accurate but not having been begun as proposed in the advertisement it cannot be considered as a claim of the premium offered … that one of them be purchased of Mr Faden to be preserved in the Society's Collection.[89]

It is not entirely clear why Faden's application was thus so firmly turned down. It may have been because the survey had taken five years to complete, considerably more than the two years stipulated by the Society, although we know that several other maps, including those of Devon and Derbyshire, breached this condition and still received awards. What seems to have concerned the Committee more was the possibility that previous surveys of the county may have been copied by Faden and not materially improved upon. They would have been very aware of the way in which Lindley and Crosley's 1792 map of Surrey had plagiarised (inaccurate) parts of Rocque's earlier map of that county.[90] The 'House of Commons' survey mentioned in the letter is probably a reference to the map produced by John Watté in 1791 entitled *A Sketch or Map of Lynn Haven and of the River Ouse to St. Mary Magdalen shewing the Proposed New Cut*,[91] together with the map prepared by Joseph Hodskinson that we have already referred to (above, p. 65), both of which were among the evidence presented to parliament in the debates over the Eau Bank cut. The reference to the Post Master General probably relates to surveys of roads undertaken with his authority; either those by John Cary or, more probably, by Daniel Paterson. Cary (who was based in the Strand, close to Faden) was commissioned in 1794 by the Post Master General in 1794 to measure the nation's post and mail-coach roads, with an initial engagement to survey a length of 9,000 miles.[92] Such a task was necessary because postage was paid, prior to the introduction of the penny postage by Rowland Hill in 1840, by the recipient, based on the mileage a letter had travelled. The results of this work were published in *Cary's Traveller's Companion*, and later in *Cary's New Itinerary*. Lt. Col. Daniel Paterson was assistant to the Quartermaster General of His Majesty's Forces and produced, between 1771 and 1826, *A New and Accurate Description of all the Direct and Principal Cross Roads of Great Britain*. In this work he was assisted by Francis Feeling, Secretary to the Post Office, acting under the authority of the Post Master General. Faden knew Paterson, having published his map of Grenada in the West Indies in 1780.[93]

In short, we do not know why Faden was passed over. It may have been because he had 'borrowed' information from elsewhere; but it may simply have been because the Society of Arts did not consider that his map was of a sufficiently high standard. Thomas Jefferys, as we have seen, seems to have had his request for a prize turned down for a very trivial reason; and Faden had, after all, already received ample recognition for other maps. It is unlikely that he was over-concerned about his failure to secure the prize for this one: and he already had other projcts well underway.

Notes

1 A Norwich physician, William Cunningham, drew a map of his home city, generally considered to be the earliest surviving printed map of any English town. Quoted in Skelton, Decorative Printed Maps, p.1.

2 L. Worms, 'Thomas Jefferys (1719–1771): beginning the world afresh', *Mapforum* 3 (Autumn 2004), pp. 20–9.

3 B. Chambers, 'Introduction', in *The County of Bedford, surveyed anno MDCCLXV, and engraved by Thomas Jefferys, Geographer to His Majesty*. Bedfordshire Historical Record Society, 1983.

4 Harley, 'Bankruptcy of Thomas Jefferys', p. 27. His bankruptcy was announced in the *London Gazette*, no. 10695, 17–20 January 1767.

5 RSA, *Society Minutes*, 24 March, 1769.

6 Will proven Somerset House P. C. C. 1771, 444 Trevor Prob 11/972 f. 224.

7 Royal Geographical Society MS, *Catalogue of Drawings and Engraved Maps, Charts and Plans; the Property of Mr Thomas Jefferys; Geographer to the King*, 1775, fol. 2, Part 1. The auction of his map stock was announced in the *Public Advertiser*, 29 January 1772. L. Worms, 'The maturing of British commercial cartography: Willam Faden (1749–1836) and the map trade', *Cartographic Journal* 41, 1 (2004), pp. 5–11.

8 A. Murphy, *An Essay on the Life and Genius of Samuel Johnson*. London, 1792, p. 133.

9 Worms, 'Maturing of British cartography', p. 8. Much of the information on the Fadens, junior and senior, has been supplied by Laurence Worms.

10 M. S. Pedley, 'Maps, war and commerce: business correspondence with the London map firm of Thomas Jefferys and William Faden', *Imago Mundi* 48 (1996), pp. 161–73. The letters are in the William L. Clements Library, University of Michigan.

11 J. Boswell, *The Life of Samuel Johnson*. Glasgow, 1823, p. 338.

12 *London Evening Post*, 29–31 October 1776.

13 M. Chrimes, 'Society of Civil Engineers', in *The Oxford Dictionary of National Biography*, Oxford University Press, on line edition, 2009.

14 A. W. Skempton and E. C. Wright, 'Early members of the Smeatonian Society of Civil Engineers', *Transactions of the Newcomen Society* 44 (1971–72), pp. 23–47.

15 Will of William Faden senior, Somerset House, P. C. C. 1783, fol. 295, Probate 11/1104/295/348 signed 24 May 1781 proved 19 June 1783.

16 L. Worms, 'William Faden', in *The Oxford Dictionary of National Biography: Missing Persons*. Oxford University Press, Oxford, 1993.

17 www.usm.maine.edu/; Harley, 'Bankruptcy of Thomas Jefferys', p. 37.

18 Worms, 'Maturing of British cartography', p. 10.

19 Harley, 'Bankruptcy of Thomas Jefferys', p. 40, although Laurence Worms (pers. comm.) thinks Jefferys had no more than five or six working on his premises.

20 Pedley, *Commerce of Cartography*, note 36, p. 276. Many of the maps in the collection are original manuscripts drawn by Officers in British Army in the Seven Year War and the American Revolution.

21 J. B. Harley, B. B. Petchenik and L. W. Towner, *Mapping the American Revolutionary War*. University of Chicago Press, Chicago, 1978, p. 1.

22 *Ibid.*, p. 84.

23 Pedley, *Maps, War and Commerce*, p. 162.

24 Harley *et al.*, *American Revolutionary War*, p. 86.

25 Withers, *Placing the Enlightenment*, p. 204.

26 Pedley, 'Maps, war and commerce', pp. 163, 169.

27   J. B. Harley, 'George Washington: map maker', *Geographical Magazine* 48 (1976), pp. 588–94. Following his death in 1799 several of Faden's maps were found in his library, including the *North American Atlas*.

28   Pedley, 'Maps, war and commerce', p. 168.

29   *Ibid.*, p. 169.

30   'Trial of Francis Henry de la Motte', *Proceedings of the Old Bailey* available in full at http://www.oldbaileyonline.org/.

31   Harley *et al.*, *American Revolutionary War*, p. 100.

32   *Ibid.*, p. 100.

33   Julian Boyd, quoted in C. Verner, 'Mr Jefferson makes a map', *Imago Mundi* 14 (1959), pp. 96–107.

34   R. B. J. Walker, 'John Russell', in *The Oxford Dictionary of National Biography*. Oxford University Press, Oxford, 2004.

35   Worms, 'Maturing of British cartography', p. 10.

36   L. Mulkream (ed.), *Thomas Pownall's Topographical Description of North America*. University of Pittsburgh Press, Pittsburgh, 1949.

37   Harley, 'English county map-making', p. 376.

38   Laxton, 'Introduction', note 93.

39   R. E. Ehrenburg, *Mapping the World: An Illustrated History of Cartography*. National Geographic Society, Washington, 2006.

40   Harley, 'The Society of Arts', p. 269.

41   The 1801 edition can be downloaded in full from www.books.google.com.

42   D. Smith, 'The Wild family firm', *Map Collector* 54 (1991), pp. 32–8.

43   Faden, *Catalogue*.

44   TNA: PRO, Prob PCC. 11/1860.232 12th April 1836.

45   TNA: PRO, Prob. 37/1125. 1838–39

46   I. Darlington and J. Howegego, *Printed Maps of London, circa 1553–1850*. George Philip, London, 1964, p. 35.

47   A. David, 'Lt. Murdoch MacKenzie and his survey of the Bristol Channel and the south coast of England', *Cartographic Journal* 40, 1 (2003), pp. 69–78.

48   Laxton, 'Introduction'.

49   G. B. G. Bull, 'Introduction', in *Thomas Milne's Land Use Map of London & Environs in 1800*. London Topographical Society publications 118 & 119, London, 1975–76.

50   G. B. G. Bull, 'Thomas Milne's Land Utilisation Map of the London area in 1800', *Geographic Journal* 122 (1956), pp. 25–30.

51   S. Foxell, *Mapping London: Making Sense of the City*. Blackdog Publishing, London, 2007, p. 188.

52   L. Dudley Stamp, *The Land of Britain: Report of the Land Utilisation Survey of Britain (1937)*. See J. E. G. Mosby, *The Land of Britain: Vol. 70, Norfolk*. Geographical Publications, London, 1938.

53   Adams, *Geometrical and Graphical Essays*, 1813 edition, pp. 302, 308.

54   W. Huddart, *Unpathed Waters: The Life and Times of Capt. Joseph Huddart F.R.S. 1741–1816*. Quiller Press, London, 1989. This book has been extensively used for details on Joseph Huddart.

55   S. Fisher, 'Joseph Huddart', in *The Oxford Dictionary of National Biography*. Oxford University Press, Oxford, 2004.

56   Terence Grocott, *Shipwrecks of the Revolutionary and Napoleonic Eras*. Caxton Editions, London, 2002, p. 111. *The Times*, 20, 21 and 24 March 1801.

57   J. W. Crerar and J. A. Ross, 'John Dalton, Captain Joseph Huddart and the Harris family', *British Journal of Ophthalmology* 37, 3 (1953), pp. 181–4.

58 J. Huddart, 'Observations on horizontal refraction which effect the appearance of terrestrial objects, and dip and depression of the horizons of the sea', paper read to the Royal Society on 24 November 1796.

59 Huddart, *Unpathed Waters*, pp. 137–67.

60 *Ibid.*, p. 111.

61 *Ibid.*, p. 112.

62 TNA: PRO, WO 78/5697 and MPH 1/ 230.

63 Huddart, *Unpathed Waters*, pp. 168–75.

64 B. Chambers, 'M. J. Armstrong in Norfolk: the progress of an eighteenth century county survey', *Geographical Journal* 130, 3 (1964), pp. 427–31. This article has been used extensively for details of the Armstrong family and their Norfolk connections.

65 Harley, 'Re-mapping', p. 63.

66 Strawhorn, 'Introduction', p. 234.

67 Gough, *British Topography*, vol. 2, p. 588.

68 R. A. Carroll, *The Printed Maps of Lincolnshire 1576–1900: Carto-bibliography, with an Appendix on Road-books 1675–1900*. Lincoln Record Society 84, Woodbridge, 1996.

69 Chambers, 'M. J. Armstrong in Norfolk', p. 427.

70 *Ibid.*, p. 428.

71 Rye, R. W., *An Index to Norfolk Topography*, Index Society, London, 1881, p. xiv.

72 Chambers, 'M. J. Armstrong in Norfolk', p. 428.

73 NRO, Colman Collection, Col/9/1, Col/9/46/5–7.

74 *Norfolk Chronicle*, 19 May 1787.

75 Chambers, 'M. J. Armstrong in Norfolk', p. 430.

76 Will dated 9th August 1791, proven 21st May 1793. NRO, N.C.C./Stills/113. Armstrong left all his goods and property to his wife Henrietta (nee Ives, from a wealthy Norwich family). There is no mention in the will of his remaining maps, surveys, or engraved plates. He wished 'to be buried in the Church of St. James, Norwich at as moderate expense as possible, to be born by six tenants, allowing them half a guinea each.' The tenants were probably from properties owned by his wife's family.

77 Norwich Millenium Library, Local Studies, *Maps,* draw 3.50.

78 Seymour, *Ordnance Survey*, p. 68.

79 TNA: PRO, WO 47/2367 f. 427.

80 TNA: PRO, WO 47/2366 f. 355.

81 *Ibid.*

82 TNA: PRO, MR 1/1170, previously WO 78/323.

83 TNA: PRO, WO 78/5697.

84 Harvey and Thorpe, *Printed Maps*, p. 111.

85 RSA, MS PR/AR/103/10/352 (hereafter RSA).

86 RSA, MS PR/AR/103/10/171.

87 RSA, MS PR/AR/103/10/312.

88 RSA, *Minutes of Committees (Polite Arts)*, 1797–1798, pp. 119–21.

89 *Ibid.*, pp. 123–4.

90 Harley, 'English county map-making', p. 375.

91 TNA: PRO, Map MP1 1/135 extracted from HO 42/151.

92 H. G. Fordham, *Studies in Carto-Bibliography*. Clarendon Press, Oxford, 1914; reprinted by Martino Publishing, Mansfield, CT, USA, 2003, p. 51 in the reprint.

93 H. G. Fordham, *Daniel Paterson: His Maps and Itineraries 1771–1825*. British Museum, London, 1925.

CHAPTER 3

The Map Redrawn

Nothing is easier than to copy former maps, ... in most of which there are numberless errors.[1]

Faden's map is still used extensively by students of Norfolk's social and agricultural history and by those with an interest in the landscape of the county. Members of the Parson Woodforde Society find that it provides them with an important spatial context for the events and scenes described in the famous diary of the Weston Longville rector, who died in 1803. The map is also important to natural scientists, who know that certain fauna and flora can only be found in places which were previously occupied by woodland or heaths. In recent years it has been employed by a range of agencies, including English Nature, the Norfolk Wildlife Trust and district and county councils undertaking 'landscape characterisation' exercises. Faden's map is thus of enduring relevance to a wide range of organisations and individuals, but in its original form it is prohibitively expensive and not readily accessible. In 1975 the Norfolk Record Society produced a good clear facsimile, accompanied by an erudite essay by the eminent historian Christopher Barringer.[2] But this, too, is now hard to obtain, and a more recent edition, produced by the Larks Press, comes in thirty-six sheets and is thus hard to use for anyone wishing to get a picture of a wide area of the county.[3] The time seemed right either to make a further copy in large sheets, or to look at the map afresh. A digital version of the map, an exact copy of the original, drawn with mapping software, opens up many possibilities. Multiple copies of the map can be printed on demand at a very reasonable cost; small parts of the map, showing a single parish or a small group of parishes, can be produced with relative ease. Above all, by using colour (earlier facsimiles were in black and white) it is possible to see the county-wide distributions of such things as parks or woods very clearly, especially as the fairly basic software employed assigns to different features their own particular 'layer', each of which can be studied individually. In essence, the digital map is like a stack of cellophane sheets laid one above the other: and it is thus possible to ask questions of the map by including certain layers but leaving out others, in order to see (for example) whether there was any spatial association between brick kilns and rivers. In addition, it is possible to add extra 'layers', not derived from the map itself, showing such things as soil types, contour lines or archaeological sites, in order to see how these relate to the features and land-use types shown by Faden. All this, it need hardly be said, ensures that a digital version of Faden's great map makes a particularly powerful tool for studying the late eighteenth-century landscape of Norfolk.

The digital redrawing was made by scanning an original version of the map, conveniently taking the form of sections of paper on linen measuring 20 by 30cm, with an A3 scanner. The software used was Adobe Photoshop, well known to photographers, and available either as a Professional or an Elements version costing £600 or £60 respectively: the cheaper version will perform almost all the tasks required. The basic drawing tools were then used to 'trace' over the original scanned map, and colour used to 'paint' some of the features: red for parks, light brown for commons, green for woodland and so on. Of the surviving copies of the original map, around two-thirds have been hand-coloured, the remainder simply being in black ink on white paper. The file type as which the image of the final map was stored, TIFF, can be used with virtually all basic imaging and drawing software.

There are some drawbacks with the digital version. Although the intention was to make an exact copy of the original, in the event it proved impossible to copy the hachures that Faden's surveyors and engravers employed to depict hills and vales. Although the original map sometimes exaggerates the magnitude of Norfolk's terrain, so that Poringland, for example, appear to sit on the top of a high plateau, the hachures are an essential element of the original map, and one of those that make it so visually attractive: the omission unfortunately serves to confirm the widespread view of outsiders that the county is 'very flat'! Bringing in a contour map as an extra layer makes a poor substitute. The software chosen has a further disadvantage: it produces a 'bitmap' or 'raster' image based on pixels which is not amenable to some forms of analysis, such as the measurement of areas. The alternative is to use 'vector'-based software, in which the lines on the map are no more than a mathematically based series of points or records. Vector maps are far easier to analyse (for such things as area calculations) and employ less computer memory. Their use does, however, require sophisticated software that is expensive, and tends to produce images that are perhaps less attractive in appearance. All these problems are likely to be resolved in the next few years as the two different types of imaging software adopt more of each other's useful properties. A further problem was presented by the fact that the original map is in six separate sheets, produced from six engraving plates; the map never existed as one single large sheet. For the purpose of the digital redrawing the six sheets have been exactly copied and are available as six separate items. But for any useful analysis of the county's landscape the six maps have been joined into a single composite sheet. This exercise has shown that the edges of the engraving plates do not always exactly coincide; occasionally a road on one sheet does not exactly join a road on the other sheet.

The principal problem with the digital version, however, is that although the surveyors used the most sophisticated methods and instruments available at the time, there are subtle distortions which ensure that the map does not fit exactly with the modern Ordnance Survey map and nor, in consequence, with data sets based on the latter, such as archaeological information or soil surveys. The digital version of Faden's map thus has to be *geo-rectified*, that is, stretched in such a way that it conforms more closely to modern surveys. This is done using specialised software (ERMapper). Fifty-three 'ground control points' were selected – positions which can be identified on both Faden's survey and on the modern Ordnance Survey, and which are unlikely to have changed position over the intervening years. Parish churches were ideal for this purpose. With these points fixed on the two maps, the software, using a quadrilateral polynomial calculation, 'warps' the Faden map to fit the Ordnance Survey. If one of the control points has been poorly surveyed or drafted on Faden's map (such as Foulsham church, which we know was wrongly positioned) the software picks this up as an anomaly, the point is rejected and an alternative found. Unfortunately, but inevitably, because the image is in raster rather than vector form, the map becomes very slightly blurred by this process, and because of this the entire map had to be drawn again prior to being analysed in relation to the various imported layers, or in terms of surface areas. The result is not perfect: the rectified map does not always coincide precisely with the Ordnance Survey. While many points are within 50m of the Ordnance Survey equivalent, some are as much as 200m adrift. In terms of analysis, however, this is less of a problem than it might sound, for many of the imported datasets, especially soil maps, are themselves not absolutely accurate and in analyzing distributions the boundaries of the latter have been 'buffered' (that is, expanded slightly) to allow for this. The inaccuracies of the geo-rectification exercise do, however, to some extent limit the uses to which the map can be put. Examining the relationship between areas and features shown on the map and archaeological sites and monuments often requires greater precision than the map can provide. Similarly, when parish boundaries are imported from the 1840 Tithe Map data, the fit with Faden's map is not always perfect.

How accurate was Faden's map?

All maps are produced for specific purposes, in this case to satisfy customers and make money for the publisher. The personal preferences and interests of the purchaser, as perceived by the surveyor and publisher, obviously resulted in some features being accentuated and others omitted. In one sense, maps are treated differently from other sources of documentary evidence by historians. If we read of a watermill in a document such as a will, we usually assume that the mill existed and ask few other questions. But if a watermill is shown on a map we immediately ask whether it been placed in the right place, whether it was still in use or long redundant, and whether another mill existed 100m further upstream which the surveyor has omitted so as not to crowd his drawing. The mill in the will is usually spared this degree of interrogation. This said, it should be obvious that the historical value of Faden's map ultimately depends on how closely it records what was actually on the ground in the late eighteenth century. It would be unreasonable to expect the degree of accuracy achieved, say, by the Ordnance Survey fifty years after its inception, when initial teething problems had been resolved. Nevertheless, if we wish to utilise Faden's map to comment on the nature of the eighteenth-century landscape we must first make some attempt to assess its veracity and the limits of its reliability. This involves considering the map from two viewpoints: in terms of its geodetic, and its topographical, accuracy.

While the exact determination of latitude and longitude was of peripheral interest to most potential purchasers of Faden's map, the accuracy of the map in this respect provides some idea of the more general care that was taken by its surveyors. Thomas Milne, as we have noted, was a recognised exponent of 'careful plotting', as suggested in particular by the fact that he was asked to provide two chapters in the standard cartographic textbook of the period. Faden's map does not have the triangulation diagrams which are drawn on some of the other county maps, such as Hodskinson's 1783 survey of Suffolk, but it may be that by the late 1790s it was thought that careful triangulation was standard practice, and that reminding purchasers that it had been employed in the production of the map was to state the obvious. Faden's map uses the meridian based on Greenwich Observatory, which his predecessor Thomas Jefferys had made standard for his own county maps. Readings of longitude and latitude are given at two points as follows:

Latitude of St Nicholas Chapel Lynn Regis	52° 45' 26"
Longitude of ditto East of Greenwich in time	0 1 36
Latitude of Yarmouth Church	52° 36' 40"
Longitude of ditto East of Greenwich	1 44 22

(the last reading is in degrees, and not in time). The map states that the Yarmouth bearings were taken by Captain Joseph Huddart by the order of Trinity House, Deptford Strond, and it is unclear whether these were verified by Milne. The readings will have been taken astronomically – that is, from the sun or stars – and it is not recorded whether further readings were taken at other points in the county. The Yarmouth church referred to is not named but it was almost certainly the parish church of St Nicholas, rather than St George's chapel. The readings show that Faden's map places the Great Yarmouth and Lynn churches within a few yards of the correct latitude, but their longitudinal positions are less accurate; Great Yarmouth church is placed 790 yards (725m) too far to the west, while the chapel at King's Lynn is 220 yards (200m) too far to the west.

When the map is geo-rectified it immediately becomes apparent that, in the original survey, the whole county was incorrectly rotated clockwise by about 3.2°. This is confirmed by a true horizontal line drawn on a photograph of the original map (Figure 27) as well as by the orientation of the map after geo-rectification (Figure 28). It is unclear how such a large error was made (presumably during drafting or engraving) and it is unlikely that the county had to be rotated to fit on the copperplate.

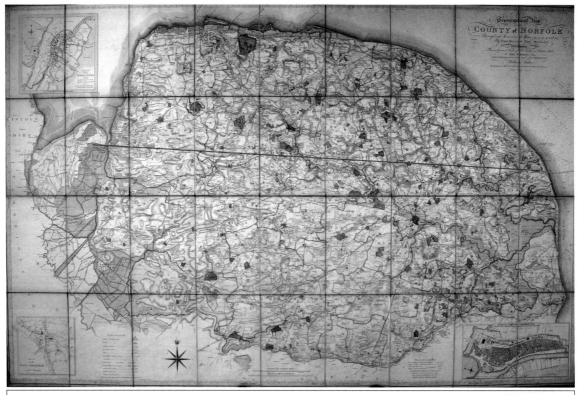

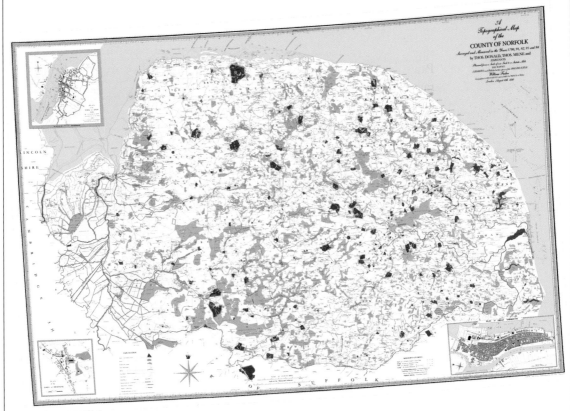

FIGURE 27. The red line represents true horizontal in July 2004 when magnetic north was 4°4' west of grid north. This shows that Faden's map appears to be rotated by 3.2° in a clockwise direction.

Figure 29 shows what happens when we compare the original map with its geo-rectified version. This confirms that the clockwise rotation is general for the whole county, and that the graticules are remarkably regular, suggesting that survey work was consistent throughout. It is not clear why the central part of the county shows slight bowing in the north–south axis, concave to the south. The original map was of course in six separate sheets, so we cannot blame this anomaly on the warping of the paper. Faden, in a small diagram inserted on the side of the map, notes that magnetic variation was 25° west in 1793. This may seem a very large variation to us, but a chart of magnetic variation over time (Figure 30) shows that the surveyors were quite correct: magnetic variation in the 1790s was indeed markedly greater than it is now. Latitude and longitude are shown on the borders of the map in degrees and minutes of time, but appear to have been added at the end of the drafting process, and not at the start. This was standard for county maps and explains why the bearings for Great Yarmouth and King's Lynn printed on the map do not exactly match the border reading.

A number of other county maps have been examined for their geodetic accuracy. David Dymond found that on Hodskinson's Suffolk map the two 'base lines' marked as 6½ miles and 7,760½ yards in length in fact measure 6.83 miles and 8,200 yards respectively.[4] These lynch-pins of the whole survey are so inaccurate that it is doubtful, in fact, whether they were the lines actually used for the survey, particularly as they were not taken across flat land with easily intervisible sites (in the case of Norfolk we do not know the site chosen for base-line measurement). Dymond also found that a straightforward comparison of distances north–south, east–west and at various places within the county suggests that true distances were underestimated by between 2.7 and 4 per cent. Paul Laxton looked at several county maps in these terms and found a wide range of variation, from the remarkable precision of Burdett's Cheshire (1777) and Greenwood's Lancashire (1818) to the disappointing map of Gloucestershire by Bryant (1824).[5] Evidently, geodetic accuracy did not necessarily improve over time, and geodetic and topographical accuracy did not go hand in hand; Peter Burdett's Cheshire may be a very fine map in trigonometric terms, but it is poorly engraved and topographically weak.

When studying Faden's map it is important to remember at all times the numerous difficulties that must have faced the surveyors and their assistants. There was no pre-existing large-scale map of the county that they could consult, although it is perhaps likely that they had access to Mostyn Armstrong's partial survey in the 1780s, and other recent surveys showing small parts of the county must also have been used. Among these were the map recently prepared by Joseph Hodskinson to accompany his report on the 'very much decayed' port of Wells on the north coast; while Faden's treatment of Lothingland, actually in north-east Suffolk, must have relied to some extent on the way that Hodskinson depicted this district on his map of that county published in 1783. Recent road surveys, like those produced by Daniel Paterson and John Cary, would have

FIGURE 28. This shows how, for the digital version, Faden's map has been rotated in a counter-clockwise direction after geo-rectification.

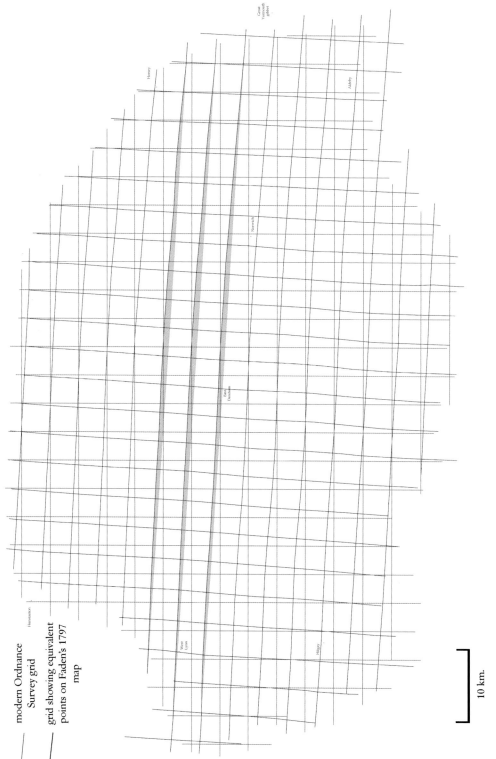

FIGURE 29. Faden's map is rotated approximately 3.2° in a clockwise direction compared to the modern OS map, as well as being slightly 'bowed' in the centre of its north–south axis, concave to the south.

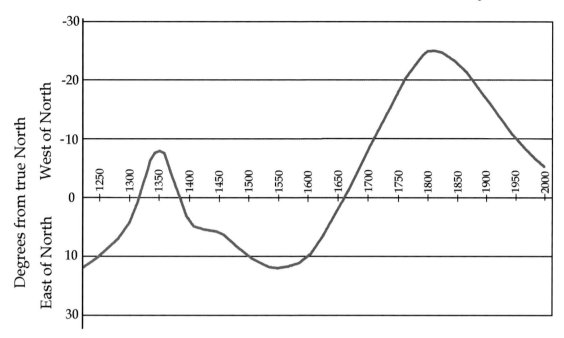

FIGURE 30. Changes in magnetic declination in England between 1250 and 2000 AD (after Clark, Tarling and Noel 1988).

been consulted, as well as maps produced by the various Turnpike Trusts in the county. Some enclosure maps would also have been available, although for the most part parliamentary enclosures in Norfolk occurred, as we have seen, in the period after the map was surveyed. Existing estate maps may have been consulted by the surveyors, and we know that in other counties the gentry were asked in newspaper advertisements to make these available. The surveyors had no legal right of access to private land and Norfolk farmers may well have regarded with suspicion men carrying theodolites, chains and plane tables attempting to cross their land while making traverse surveys. We can calculate from the number of years involved in surveying the county and the number of parishes involved that only three and a half days could have been spent, on average, surveying each. We have no idea of how many assistants were available to Donald and Milne but from what we know of similar work in other counties it is unlikely that they numbered more than two or three.

It is not possible to list all of the map's various inaccuracies, as this would require an intimate knowledge of the topography and history of every parish in the county. Suffice it to say that, while undertaking the digital redrawing, many errors, mainly of a minor character but some major, became apparent; and more were reported by people who have used the redrawn map. These mistakes could have been made at a number of stages: by the surveyor, the draughtsman, the engraver – and of course in the process of copying and drawing the final computer-based image. The principal mistakes are listed in Appendix 5 but include the omission of some churches (including those at Haveringland, Waxham, Shipdham, Pulham Magna (modern Pulham Market), Carleton,

Sporle, Stow Bardolph, West Ruston (Ryston), Wiggenhall St Germans, Thorpe St Andrew and others) and the transposition of some neighbouring churches (Surlingham St Saviour and St Marys, East and West Runton). Where more than one church stands within a churchyard, an idiosyncrasy of the Norfolk landscape, only one is usually shown (as at Reepham). A number of major country houses are not marked, including Barnham Broom Hall, Thurgarton Hall, Thorpland Hall, Shelton Hall, Sedgeford Hall, Flordon Hall, Fincham Hall and Gonville Hall near Wymondham. More worryingly, there appear to be some 'black holes' on the map, into which one suspects the surveyors did not venture, so that a whole series of features are omitted. One lies just to the north of Flordon village; another in the area between Hedenham and Tindall woods. In the latter case two substantial ancient woods – Long Row and Round Grove – were completely left off the map, as was a small settlement. In general, Faden is poor on woods. Sexton's Wood in Hedenham has unquestionably been in existence for centuries – it was owned by the Bigod family, who were based in nearby Ditchingham from the 1270s.[6] Surviving records suggest that in medieval times the wood covered some 86 acres, and it is shown with this area on Bryant's map of 1826, and on the first edition of the one-inch Ordnance Survey. Faden, however, shows a wood covering over 160 acres, and with an outline quite different to that depicted on all other maps. Other errors include the incorrect positioning of Stanfield Hall near Wymondham, and the depiction of rivers (a tributary of the Thet between Banham and Eccles, and the upper Wensum between Great Ryburgh and Raynham) in such a way that they appear to flow in two directions! Hundred boundaries in general are inaccurately drawn and this may reflect the fact that they were to be virtually abandoned as a useful administrative tool within the following four decades.[7]

Accuracy: eleven parishes analysed

The accuracy of Faden's map is best assessed by examining in some detail a relatively large block of countryside. There is of course a risk in such a procedure. The chosen area may have been surveyed by only one of the two surveyors, or by their less experienced assistants: different individuals clearly had different perspectives on the landscape, so that certain features – such as inns – are recorded in greater detail in some areas of the county than in others. For this exercise, therefore, a substantial block of territory was examined, comprising eleven contiguous parishes in north-central Norfolk (Figure 31): Guist, Swanton Novers, Foxley, Croxton with Fulmondeston, Wood Norton, Hindolveston, Great and Little Ryburgh, Stibbard, Foulsham and North Elmham, which together cover an area of 9,480 hectares, 4.5 per cent of the total land area of the county. Faden's depiction of this area can be compared with how it is represented on various enclosure maps, on Bryant's 1826 map of the county (which is at roughly the same scale as Faden's map), on the Old Series Ordnance Survey one-inch map (surveyed 1815–18 and published in

Norfolk
The eleven parishes
studied

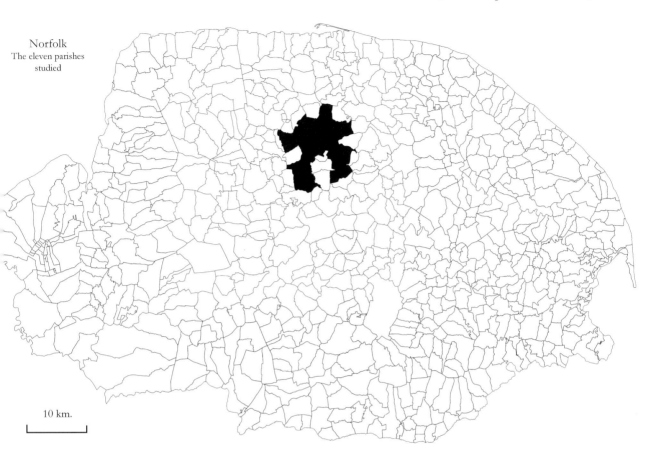

10 km.

FIGURE 31. The eleven
parishes in central-
north Norfolk analysed
in detail to assess the
accuracy of Faden's map
(see text).

1836), and on the modern Ordnance Survey. Comparison with local enclosure
maps is particularly instructive, as these were produced within fifteen years of
Faden's map (except in the case of North Elmham, which was surveyed in 1829),
at a large scale, and of necessity to a high degree of accuracy (the maps were
moreover on display to the public, often in a local hostelry, and errors would
soon have been pointed out). A full analysis of the results of this procedure are
presented (in the form of maps and spreadsheets) on the DVD accompanying
this volume: what follows is a brief summary.

Foulsham is one of very few parishes in Norfolk where the surviving enclosure
documents include 'draft' maps, showing the landscape immediately before the
enclosure took place (for most parishes we have only the map accompanying
the enclosure award, showing the disposition of the new properties, making
comparison with Faden's map more difficult).[8] Here we see that the commons,
heaths and greens are very close to those on Faden's map, surveyed some twenty
years before. The layout of roads and tracks and the disposition of farms and
cottages, both in the village centre and on the edges of commons, is likewise
closely comparable. The church is misplaced by about 150m but the newly
erected (1783) parish workhouse is correctly positioned. Faden seems to show

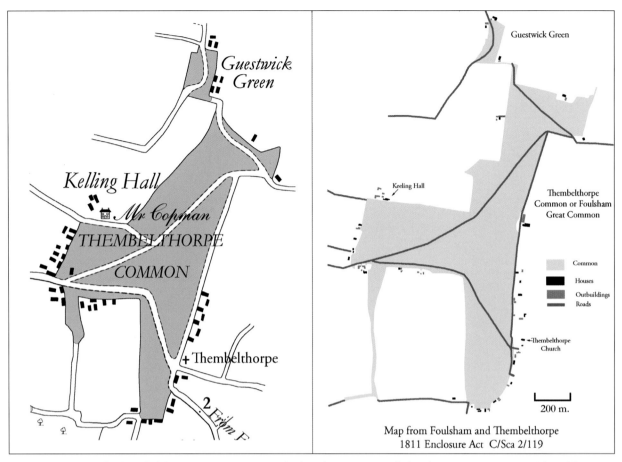

Map from Foulsham and Thembelthorpe
1811 Enclosure Act C/Sca 2/119

the shape of Thembelthorpe Common, and the disposition of the houses around it, quite accurately, although the north-western edge of the common is straightened and oversimplified (Figure 32). Faden does, however, show a small, rather intriguing piece of woodland to the south of the village, for which there is no other evidence. Whether this had been removed by the time the draft enclosure map was surveyed, or is a fiction, remains unclear.

There is no pre-enclosure map for Swanton Novers but some useful comparisons can be made with the enclosure award map.[9] The disposition of houses and roads is very similar on both maps. Although the outlines of the three large woods in the parish are rather simplified on Faden the woods themselves are placed in their correct positions. We can, in fact, compare the areas of these woods on four different maps: Faden (1797), the enclosure map (1811), Bryant (1826) and the modern Ordnance Survey. This suggests that Faden underestimated their size, although not by any very great amount. In this case – and in other examples – we need to remember that woods are notoriously difficult to survey accurately as it is hard, if not impossible, to obtain a clear line of sight through them. Bryant, it should be noted, has simply copied Faden's rather poor drawing of the 'Great Wood', a practice apparent elsewhere on his map.

FIGURE 32. Themelthorpe common, Foulsham, as depicted on Faden's map of 1797 *(left)* and the 1811 enclosure map *(right)*.

In Foxley, the depiction of roads, trackways and settlements is generally very similar on Faden and the enclosure map.[10] The parish includes Foxley Wood, the largest area of ancient woodland in Norfolk, and all the maps examined show it with the same general outline. Its area is shown as 107 hectares on Faden's map, 117 hectares on the enclosure map, 121 on the First Series OS and 126 on the modern Ordnance Survey. The increase in area between the 1790s and the time of the enclosure in 1814 may be explained by a slight expansion to the south-west.

As already noted, a full comparison of the way that the other eight parishes examined in this exercise are shown by Faden, and on later maps, can be found on the accompanying DVD, but overall we can summarise the accuracy of Faden's map as follows. Churches, somewhat surprisingly, are not always recorded correctly (Elmham, Foulsham and Stibbard churches are all slightly misplaced while in Wood Norton, St Peter's church – probably by then already in ruins – is in completely the wrong place). Roads, lanes and trackways are generally shown with a high degree of accuracy, although the winding course of some examples has been simplified and straightened: where errors occur they tend to come in batches, as in the area to the south of Stibbard village (although in the absence of earlier maps it is unclear whether the surveyors made a series of errors here, or whether there were indeed major changes to the local road network between 1790 and 1810). Woodland is, on the whole, poorly shown on Faden's map, as Oliver Rackham and others have pointed out: it was not considered a reliable source when the Inventory of Ancient Woodland was compiled.[11] This deficiency is shared by many of the county maps. Thus Phillips, in his analysis of William Yates's map of Staffordshire, noted that woodland covered only 2.4 per cent of the county and yet this was a county predominantly already enclosed and 'well stocked with timber'.[12] In the case of Faden, however, caution should not perhaps be taken too far. Some of the larger woods in the county are accurately represented, in terms of both shape and size: and we should note that the Ordnance Survey first series does not come out very well in this respect either, with the county's larger woods generally being too small by between 5 and 10 per cent. In these eleven parishes, at least, there is no evidence that areas of existing ancient woodland were actually omitted by Faden's surveyors. Faden's representation of rivers appears to be particularly detailed and generally accurate, a feature of the map which perhaps relates to one of the conditions for the Society of Arts prize. The Society demanded precision in this respect, as well as the provision of correct 'levels', a requirement associated with the contemporary interest in extending the system of navigable inland waterways.

Faden's depiction of commons, heaths, greens and other 'waste' is, as we shall see, of particular interest. Parliamentary enclosure was well underway in some parts of the county when Faden's surveyors were at work, and the principal purchasers of the map, the landed gentry, would thus have been particularly interested in the location, disposition and extent of what they considered to be unproductive, but potentially improvable, land. For only two of the sample parishes do pre-enclosure maps survive, allowing us to directly check the veracity of Faden's depiction of commons and heaths: but in most of the others in the sample the 'ghosts' of their outlines can be seen in the disposition of the houses and cottages which had formerly fronted on them. In general, areas of 'waste' (heaths, commons and warrens) on Faden appear much as on the enclosure maps, although with their outlines sometimes simplified. Faden also seems to present their area accurately, although in general the smaller commons are drawn slightly larger than they should be. There are also some problems with what precisely Faden included, and excluded, in his category of 'commons and wastes', for the map also marked areas of 'marsh and fen', many of which were undoubtedly also common land.

The way in which the map represents settlement is also of some interest. Houses are shown schematically, as black rectangles, but given the scale of the map it would have been impossible to represent every individual building: Faden thus presents us with a simplified version of the settlement pattern (even on the modern equivalent 1:50,000 maps it is groups of buildings, rather than individual houses, which are shown). In smaller villages the number of 'blocks' generally does correspond fairly well with the number of buildings shown on the more detailed of the enclosure maps. But in the case of larger settlements there were simply too many for this to have been possible. In general, small hamlets and individual farms are thus very well represented on the map, with only a handful of exceptions (such as Manor Farm in Wood Norton). Settlements strung out along common edges – a very important type of settlement in late eighteenth-century Norfolk – are also generally depicted with a reasonable degree of accuracy.

In the eleven parishes examined Faden seems good on watermills. Elmham mill is mistakenly called Bintry mill and at first sight its description as a 'paper mill' also seems a mistake: but in fact Faden is correct here – it was run for about twenty years as a papermill before the business failed. On the whole, if Faden shows a feature it was probably there, or thereabouts; but this does not mean that some things were not omitted. On Faden's map, as to some extent on all maps, absence of evidence cannot be taken as evidence of absence.

As far as we are aware, only a handful of the county maps have been analysed from the perspective of landscape history. Harley examined the 'wasteland' (heaths, commons, mosses and marshland) shown on Burdett's map of Cheshire and Yates' of Lancashire, as well as the distribution of windmills and watermills.[13] The county maps were also used in the 1930s in some of the volumes produced by the Land Utilisation Survey, to provide some comparison with contemporary patterns of land use. In the case of Norfolk John Mosby painstakingly traced commons, woodland, marshes and estates on Faden's map and used a photographic reduction technique to produce a map showing the distribution of the main kinds of land use in the county in the late eighteenth century.[14] The Cheshire and Lancashire maps were similarly used by Dudley Stamp to show that the area of 'moss' in Lancashire fell from 34,500 acres in the late eighteenth century to a mere 4,500 acres in 1931–2.[15]

Norfolk, Faden and the Board of Ordnance

In trying to place Faden's Norfolk map in its wider historical context we must lastly consider the activities and influence of the Board of Ordnance before, during and after the period in which the map was surveyed and published. The period from 1785 to 1800 witnessed the birth of 'official' mapping in Britain, and was a period of intense scientific endeavour which could not have failed to influence those involved in producing the Norfolk map. During this time there was much mutual exchange of both ideas and personnel between county map-makers and the fledgling map department in the Board of Ordnance. By examining briefly the twenty-five years following the publication of Faden's map we can thus learn a little about how well, or how poorly, the Norfolk map-makers performed.

In seeking the origins of the 'official' mapping of what was to become the United Kingdom we must first note the influence of William Roy, originally from Lanarkshire, who was initially involved in the 1747–55 military survey of Scotland which followed hard on the heels of the 1745 Jacobite rebellion. He was well aware that the available maps of Britain were greatly inferior to those of Austria, Denmark, France or even India – and also aware of the potential military implications of this deficiency. He was critical of existing county maps, complaining to George III that:

> These County Maps are sufficiently exact, in what regards their geometrical measurement, for common purposes, but are extremely defective with respect to the topographical representation of the ground, giving scarcely any Idea, or at least but a very imperfect one, of what is remarkably strong or weak in the nature of the Country.[16]

By 'strong' and 'weak' he was referring to relief, a factor of much importance in the construction of fortifications and the movement of troops. In the early 1760s he formulated a plan for producing a triangulation of the whole country, into which his existing map of Scotland could be integrated. He presented a paper entitled 'Considerations on the Propriety of making a General Military Map of England, with the Method proposed for carrying it into Execution, & and Estimate of the Expense'. He estimated that such a project would cost £2,778 in the first year and £20,000 in total – not dissimilar to the cost of the *Carte de France*.[17] The sums alarmed the government, however, and his ideas were dropped: it was decided to leave the mapping of the country, as we have seen, to private enterprise.

The American War delayed progress in plans for a countrywide survey, but in 1783 Cassini de Thury, a foreign member of the Royal Society, suggested that France and England should cooperate for a change, and try in particular to resolve a difference of opinion concerning the relative positions of the Royal Observatory in Paris and the Royal Observatory at Greenwich. The French scientists claimed that Greenwich was incorrectly recorded by 11 seconds of longitude and 15 seconds of latitude.[18] France, of course, had already completed much of its nationwide triangulation, certainly as far as Calais: England had not even begun. William Roy, who was also a member of the Royal Society, was asked to respond to the challenge and he immediately saw that this gave him an opportunity to rekindle his earlier plans. By now a Major General in the Engineers, he used soldiers to clear a base line on Hounslow Heath that would be the cornerstone of a nationwide triangulation. The exact measurement of this base line was so important that a series of experiments were carried out using at first 100-foot chains, then 20-foot seasoned Red Riga deal rods, and finally 20-foot rods of one-inch glass tubing, in order to ascertain which would give the most accurate results under varying temperatures. The measurements, undertaken in the autumn of 1784, took thirteen days and although carried out with scientific purpose were accompanied by an almost carnival atmosphere, with tents and liquid refreshments. The event was attended by George III and Sir Joseph Banks, President of the Royal Society, as well as by John Smeaton and Jesse Ramsden, both of whom were well known to William Faden. The final computation, on 30 August, gave a length of 27,404.01 feet at 62° Fahrenheit.[19] Ramsden was asked to build a theodolite able to continue the triangulation exercise from Hampton Poor House and the King's Arbour, which stood at the ends of the base line, on into the rest of the country. Renowned as a perfectionist, he took three years to complete the instrument, to the increasing frustration of William Roy. He was also renowned for his other-worldliness and Richard Lovell Edgeworth recorded a famous episode when Ramsden drove down to Kew Palace with an instrument which George III had ordered. He asked if the king was at home, and insisted on being admitted. The king received him graciously and, after examining the instrument said to him, 'I have been told, Mr Ramsden, that you are considered to be the least punctual of any man in England; you have brought home this instrument on the very day that was appointed. You have only mistaken the year.'[20]

When finally constructed, at a cost of £373 14s, his famous 'Great Theodolite', or 'Great Circular Instrument' – aptly named, given that it weighed 200lbs – could read a mark at a distance of seventy miles with an angular error of no more than two seconds. It continued to be used until 1853, only eventually being destroyed by enemy bombing of Southampton in 1940.[21] With the theodolite complete, the triangulation exercise was continued down to the English Channel at Dover Castle and Fairlight Head; and then across to Cape Blanc Nez and Mont Lambert, north of Boulogne in northern France, which was reached in October 1787. The results of the Greenwich to Paris triangulation were finally published in 1789 in the *Philosophical Transactions*, sadly three weeks after Roy's death.[22]

The Trigonometrical Survey was inaugurated in 1791; thus a civilian project that had started under the auspices of the Royal Society came under the umbrella of the military in the form of the Board of Ordnance. The first directors were Lt. William Mudge of the Royal Artillery (a friend of William Faden through the Smeatonian Society) and Isaac Dalby, a civilian. One of their first tasks was to repeat the measurement of the Hounslow Heath base line, producing a result which differed from the original by 2¾ inches. It was decided to take an average of the two readings at 27,404.24 feet.[23] In 1794 a verification base line was laid out on Salisbury plain and found to be only a couple of inches different from the length calculated by triangulation. With the threat of war with France looming, the Survey concentrated on completing the south coast and their results were published regularly between 1799 and 1811 (by William Faden 'at his private expense') and made available to private county map surveyors.[24]

The topographical survey to 'fill up the triangles', referred to as the 'Internal' or 'Interior' Survey, proceeded alongside the Triangulation Survey, although it had started a little earlier, in 1787. It had different personnel and was under the direction of William Gardner (who had worked as surveyor on the Sussex county map between 1778 and 1783). In general, the standard of topographical surveying prior to 1820 was well below that of the Trigonometrical Survey, and the early Ordnance maps were recognised by contemporaries as falling well short of an acceptable standard.[25] In July 1821 William Faden received a letter concerning one of them:

> The mounting is so inaccurately done that the bottom from Edge to Edge is one and a quarter inches wider than the top and the left or Western side of the canvas is one and three-eighths inches longer than the opposite side, consequently the Angles at the bottom are not right angles and it is impossible to hang it so as not to offend a moderately accurate eye[26]

By the 1820s the Board of Ordnance was well aware that the public lacked confidence in their maps. Indeed, this was the main reason for the continuing success of the county maps, which were now being produced by Andrew Bryant and the Greenwood brothers. William Mudge of the Board of Ordnance was even included in the subscription list for a proposed private map of Lancashire which suggests a certain lack of confidence in his own organisation.[27] The Trigonometrical Survey was truly at the frontier of geodetic science, but the topographical side of the enterprise continued with practices which were, in many ways, little better than those used in the more mediocre of the county surveys. This was in part because some of the work was carried out by unsupervised civilians, and by cadets training for a Commission in the Corps of the Royal Engineers.[28] Kent was the first county surveyed under the aegis of the Board of Ordnance but, having no experience of publication, they turned to a private map-maker, William Faden, to engrave and publish the resulting map in 1801 (it was in fact ready for publication in 1799 but, possibly for military security reasons, it did not see the light of day for a further two years). On several occasions Faden had made himself invaluable to the Board, and it is understandable that they should have chosen him, rather than John Cary or one of the other London map publishers. Because of security concerns the engraving was carried out under Faden's guidance in the Tower of London, rather than in his Charing Cross premises. The engraver was Thomas Foot, working freelance from Weston Place, St Pancras; he had worked previously for Faden on the second edition of the Sussex county map. In 1805 Faden was given exclusive rights to sell the early Ordnance Survey maps. Other London map-sellers made their unhappiness at being excluded from this business opportunity very apparent; Faden for his part complained to the Board of Ordnance that the profit they allowed him (a 17 per cent discount and six months' credit)[29] was well below what he was accustomed to receive from the sale of private county maps.[30] He received some recompense, however, from the lucrative contract he was given for mounting and boxing the large number of presentation copies made for the Royal Family and other notables, including members of the Cabinet, the universities and senior officers in the Board of Ordnance and Quartermaster General's Department.[31]

In 1799 the thoughts of the Board of Ordnance turned to their next map, that of Essex. The Kent survey had taken longer than anticipated, and had also cost more, the total bill coming to no less than £8,000.[32] This may partly be explained by the decision to survey parts of the map at a scale of three inches and even six inches to the mile, for military reasons. The Board was now under pressure to produce a cheaper map. The final cost estimates for Essex, surveyed at two inches to the mile but printed at one inch, were remarkably similar to the expenses incurred in producing Faden's map of Norfolk some five years earlier.

> The County of Essex contains 1563 square miles of which number 1250 remain to be surveyed [some had already been surveyed as part of the Kent map]
>
> | | |
> |---|---:|
> | to Surveying 1250 square miles at 33 shillings per square mile | £ 2062 0 0 |
> | to Carrying on the Trigonmetrical Survey of the County | £ 300 0 0 |
> | to travelling charges of Ten Draftsmen | £ 60 0 0 |
> | Total | £ 2422 0 0[33] |

The early history of the Ordnance Survey in Norfolk is somewhat complex and needs to be considered under two separate headings, trigonometrical and topographic.[34] The trigonometrical survey for Norfolk was considerably delayed because of its flat terrain; King's Lynn and Wisbech had been covered by 1811 (Figure 33) but the rest of the county was included only between 1820 and 1824.[35] One of the problems was that the fixed poles, erected for surveying, were repeatedly pulled down by local people, and required protection by the Justices of the Peace.[36] The topographical survey, moreover, proceeded without the trigonometrical data, with results that could have been foreseen; much of the county had to be resurveyed over the subsequent twenty years (Figure 34). The original drawings for the first Ordnance Survey, drawn at two inches to the mile, are kept in the British Library; they vary greatly in quality, 'ranging from scribbles to works of art'.[37] Thomas Colby took over from William Mudge as Superintendent of the Ordnance Survey in 1820 and soon became aware that Norfolk had been very inadequately surveyed. He wrote:

FIGURE 33. The principal triangulation points in England and Wales used in 1809 (from Mudge and Colby Account of the Trigonometrical Survey, 1811). From Harley and O'Donoghue, Old Series Ordnance Survey Maps.

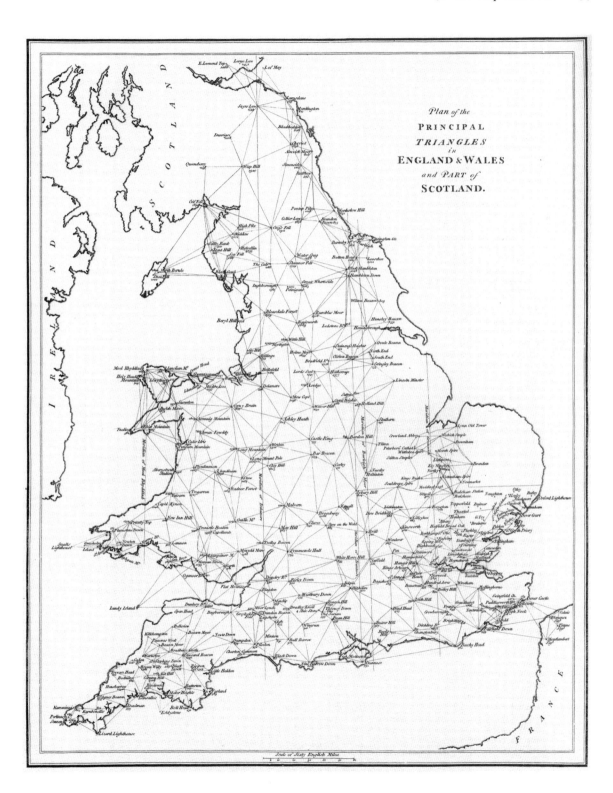

Plan of the
PRINCIPAL
TRIANGLES
in
ENGLAND & WALES
and PART *of*
SCOTLAND.

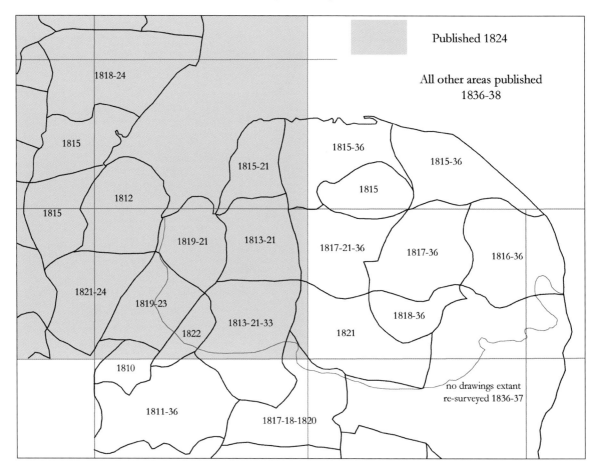

The Lynn plan of Mr Yeakell Junr. [the son of Thomas Yeakell, the Sussex county map surveyor and later Chief Draftsman in the drawing room of the Tower of London] is done in a most slovenly inaccurate manner; one wood was fully double its real size and more than twice its breadth … in short there was too much sketching and that of a very bad quality … I did not think we could in the course of two or three hours have discovered so many errors in any plan.[38]

FIGURE 34. Dates of the Ordnance Survey drawings and revisions in Norfolk (after Harley 1987 and Hodson 1989).

Colby immediately instituted reforms and replaced civilians, paid by the number of acres they surveyed, a form of piece-rate, by military surveyors. He wrote that 'surveyors who had performed grossly inaccurate plans should be charged for their correction'.[39] He was able to publish the western section of the county in 1824 but the rest, as noted, had to be extensively resurveyed in the 1830s, delaying final publication until 1836–8. Uniquely, the earlier publication was helped by private Lincolnshire subscribers, motivated by their fox-hunting interests, each paying (via William Faden, needless to say) four and a half guineas.[40]

This short digression, covering a period extending for some period thirty years after the publication of Faden's map, indicates how difficult it must have been to survey a relatively flat county like Norfolk: and how well Donald and Milne did their work, with resources which were much inferior to those available to the Ordnance Survey.

Conclusion

Mark Monmonier has stated that 'not only is it easy to lie with maps, it is essential'.[41] How else can one compress a three-dimensional space onto a two-dimensional surface at a highly reduced scale? Compromises had to be made to produce Faden's map of Norfolk: by the original surveyors, the compiler, the engraver and the publisher; and others have been made in producing this new, digitised version. At each step, choices have had to be made about what should be included and, more importantly, what should be left out; and at each step errors made earlier in the process have been compounded. These caveats must be borne in mind through all the pages that follow. We must also be wary of looking at the county maps only through modern eyes and remember that the opinions of contemporaries may be just as valid. In 1780 Richard Gough made a note concerning Peter Burdett's 1767 map of Derbyshire: 'Mr Michell s^d of it, y^t was ye best English map we have by much'.[42] With hindsight we can easily criticise the Derbyshire map for its topographical failings,[43] but the map was geodetically outstanding and was duly awarded a prize by the Society of Arts.

At the start of this book we said that we would try to place Faden's Norfolk in the context of the 'cartographic revolution' represented by the eighteenth-century county maps. But with so many of these maps still awaiting serious analysis, what we have offered can only be a very preliminary view. Exactly 100 years ago Herbert George Fordham, the 'father of British cartography', wrote of the county maps that:

> The whole subject merits a careful and comparative study which up to the present seems never to have been attempted. To do it justice the publication of a considerable number of facsimile reproductions would be essential.[44]

As we have seen, copies of these maps, of variable quality, have indeed been published but their analysis has tended to concentrate on the history of production and on the personalities involved, rather than on the map itself and what it can tell us about landscape and society. What we can say, and with some confidence, is that Faden's map of Norfolk would stand at or close to the top of any ranking of these maps, whatever the criteria adopted. Some of them, accurate in geodetic terms, are let down by poor engraving. Some, beautifully drawn, are marred by topographical inaccuracies. Faden's map of Norfolk, in part perhaps because it was produced so late in the series, is both remarkably accurate in geodetic and topographic terms, *and* pleasing to the eye. The county was served by three of the best surveyors and hydrographers in the country, and the whole operation was overseen by a publisher of quite extraordinary energy and business acumen.

Notes

1   'Editorial', *Gentleman's Magazine* 17 (1747), p. 406.

2   J. C. Barringer, 'Introduction', in *Faden's Map of Norfolk*. Norfolk Record Society 42 (1973), pp. 1–16.

3   Barringer, 'Introduction', 1989, pp. 1–11.

4   D. P. Dymond, 'Introduction', in *Joseph Hodskinson's Map of Suffolk in 1783*. Larks Press, Dereham, 2003.

5   Laxton, 'Geodetic and Topographical Evaluation', p. 42.

6   O. Rackham, 'The ancient woods of Norfolk', *Transactions of the Norfolk and Norwich Naturalists' Society* 27, 3 (1986), pp. 161–77.

7   The hundred boundaries on Faden are so inaccurate that their boundaries, as well as those of parishes, have been imported as a separate 'layer' to allow analysis of the map in later chapters. This layer is based on R. Kain and R. Oliver's *Historic Parishes of England and Wales: An Electronic Map of Boundaries before 1850 with a Gazetteer and Metadata*, History Data Service, Colchester, 2001. This data was essentially collected from tithe maps.

8   NRO, Enclosure of Foulsham and Themelthorpe in 1811, C/Sca 2/119, BR 25/1–11, 20–22, 39. MC 662.

9   NRO, Enclosure of Hindolveston, Swanton Novers and Wood Norton in 1811, C/Sca 2/152.

10 NRO, Enclosure of Foxley in 1814, C/Sca 2/120.

11 J. Spencer and R. Thomas, *Norfolk Inventory of Ancient Woodland*. English Nature, Peterborough, 1992.

12 Phillips, 'Introduction', p. xiii.

13 Harley, 'William Yates and Peter Burdett'.

14 Mosby, *Norfolk*.

15 W. Smith, *The Land of Britain: Vol. 45, Lancashire*. Geographical Publications, London, 1941, p. 70.

16 J. Fortescue, *The Correspondence of King George the Third from 1760 to December 1783*. London, Macmillan, 1927, vol. 1, p. 330.

17 Seymour, *Ordnance Survey*, p. 8.

18 *Ibid.*, p. 14.

19 W. Roy, 'An account of the measurement of a base on Hounslow Heath', *Philosophical Transactions* 85, 2 (1785), p. 477.

20 R. L. Edgworth, *Memoirs*, 1, pp. 191–2 (1820).

21 Jane Insley, 'The tale of the great theodolites', paper delivered at FIG conference, Stockholm, 2008: www.fig.net/pub/fig2008/papers/hs03/hs03_01_insley_2838.pdf

22 Close, *Ordnance Survey*, p. 24.

23 H. G. Fordham, *Some Notable Surveyors and Map Makers of the Sixteenth, Seventeenth and Eighteenth Centuries and their Work: A Study in the History of Cartography*. Cambridge University Press, Cambridge, 1929.

24 Mudge, *Accomplishing a Trigonometrical Survey*.

25 J. B. Harley, 'Error and revision in early Ordnance Survey maps', *Cartographic Journal* 5 (1968), p. 114.

26 Ordnance Survey letter book ff.339–340, quoted in Seymour, *Ordnance Survey*, p. 58

27 Harley, 'Error and revision', p. 117.

28 Seymour, *Ordnance Survey*, p. 50.

29 Pedley, *Commerce of Cartography*, p. 196.

30 Seymour, *Ordnance Survey*, p. 74.

31 *Ibid.*

32 J. B. Harley and Y. O'Donoghue, 'Introduction', in *The Old Series Ordnance Survey Maps of England and Wales, Vol. 1, Essex, Kent, E. Sussex, and S. Suffolk*. Harry Margary, Lympne Castle, 1975, p. xxix.

33 *Ibid*, p.xxix.

34 J. B. Harley, 'Introductory essay', in *The Old Series Ordnance Survey Maps of England and Wales, Vol. 5, Lincolnshire, Rutland, and East Anglia*. Harry Margary, Lympne Castle, 1987.

35 J. B. Harley, 'Cartographical introduction', in series *Reprints of the First Edition of the One-inch Ordnance Survey of England and Wales*, Norfolk sheets 37, 38, 45, 46 and 47. David and Charles, Newton Abbot, 1970.

36 *Ibid.*, sheet 46.

37 Y. Hodson, *Ordnance Surveyors' Drawings, 1789–c.1840*. Research Publications, Reading, 1989.

38 Ordnance Survey Letter Book, ff. 122, 123, quoted in Seymour, *Ordnance Survey*, p. 101.

39 Harley, 'Cartographical introduction', sheets, 45, 38 and 46.

40 Harley, 'Introductory essay', p. viii.

41 M. Monmonier, *How to Lie with Maps*. University of Chicago Press, Chicago, 1991, p. 1.

42 BLO, Gough Gen. Top. 361, Vol. 1, fol. 280. This probably refers to John Michell (1724–1793), the Cambridge mathematician and scientist who could have been referring to the latitudes and longitudes in the map.

43 Harley *et al.*, 'Introduction'.

44 H. G. Fordham, 'Notes on the cartography of England and Wales', paper read at a meeting of the British Association in Dublin, September 1908.

CHAPTER 4

Commons, Greens and Heaths

....

Commons and wastes

A common was an area of land owned by an individual – usually the lord of the manor within which it lay – but which certain other people had the right to use for grazing, and for gathering fuel and other useful materials. The most immediately striking feature of Faden's map of Norfolk is the sheer number, and extent, of commons that it shows still remaining in the landscape in the late eighteenth century. The attention the map pays to these features reflects the interests of the landowners and larger farmers who formed an important part of Faden's potential market, for this was a time when rising grain prices, coupled with a fashionable interest in agricultural 'improvement', ensured that there was a widespread interest in the 'enclosure' of such areas – their division between those who had rights over them, and their conversion to private property.[1] Indeed, commons appear to have been surveyed more accurately than almost any other feature of Norfolk's landscape – more accurately, for example, than landscape parks, and much more accurately than woodland – although the fact that commons were open and in effect public spaces, which could thus be surveyed (as it were) from the inside, may also have been a factor in this.

The county-wide depiction of common land is perhaps the map's greatest contribution to our understanding of landscape history, for the overwhelming majority of these areas were to disappear over the next three decades or so, usually through enclosure by parliamentary act. Sometimes this took place at the same time as the common fields in a parish were removed, but often these had disappeared much earlier, as strips were consolidated and enclosed in a gradual, piecemeal way, leaving only the commons (hard if not impossible to enclose by such means) to be removed by more formal methods of enclosure.[2] As we have already noted, parliamentary enclosure, which affected between a fifth and a quarter of England's land area, was largely concentrated in two great waves. The first, peaking in the late 1770s, was mainly concerned with the enclosure of open fields on the Midland clays of England, and their conversion to pasture. The second, which occurred during the high price years of the Napoleonic Wars, tended to affect light arable land, especially in eastern England, but was mainly concerned with tracts of common grazing, especially heaths and upland moors.[3] Most of Norfolk's commons were enclosed during, and immediately after, the Napoleonic Wars, and Faden's map thus shows them on the very eve of their destruction.[4]

While most parliamentary enclosures of common land thus took place in the period after the map was surveyed a few had occurred earlier (the earliest was the enclosure of the marshes at Stokesby in Flegg in 1722), and Faden himself notes several places where commons had recently disappeared. The words 'Common inclosed' thus appear to the west of Old Buckenham, a parish subject to an enclosure award of 1790, while 'Thelton Common inclosed' are written across the area covered by the parish of Thelveton. There was no enclosure act for this place so presumably the common here, which may have extended into the adjoining parish of Thorpe Parva, was enclosed by some other means (probably by unity of possession: both were small, 'closed' parishes under the control of single landowners). Similar comments appear with regard to Carleton Rode Common (enclosed by parliamentary act in 1777) and Banham Common (enclosed by an act of 1789). The map also notes 'Drayton Heath Lately Enclosed', which is something of a problem as the parish was not, in fact enclosed until 1811. In all, forty-six enclosures by parliamentary act were made in Norfolk in the period before Faden's map was surveyed (1722–1790).[5] Several of these dealt with more than one parish, so that no less than fifty-five places may have lost their commons before the map was made. Moreover, a number of places will have been enclosed without recourse to parliamentary act, as the example of Thelveton indicates, by the simple expedient of buying up all the commonable tenements. And in addition to all this, as we shall see, commons could be, and frequently were, partially enclosed and encroached upon in a variety of ways in the course of the sixteenth, seventeenth and eighteenth centuries.

What Faden shows us is thus a snap-shot – the extent and configuration of common land as it was in the 1790s – and not the totality of commons as they would have been in the sixteenth or seventeenth centuries, still less in the Middle Ages. On the other hand, maps from the sixteenth and seventeenth centuries generally show commons with much the same boundaries as those depicted by Faden. Moreover, for reasons which remain unclear, a number of commons are shown on the map in spite of the fact that the parishes in which they lay had been enclosed by parliamentary acts some time *before* the 1790s. These were mainly in the west of the county (principally in the parishes of Brancaster (enclosed 1755), Titchwell (1786), Ringstead (1781), Heacham (1780), Dersingham (1779), Great Bircham (1739), Tottington (1774), Weeting and Broomhill (1774) and Grimston (1780)), although there were also examples in the centre (Hingham (1781), Litcham (1760), Scarning (1765), Beetley (1774), Swanton Morley (1755), North Tuddenham (1764) and Little Ellingham (1766)), as well as scattered examples elsewhere (Walpole St Andrew (1789); Felthorpe (1777)).[6] The survival of these commons may be more apparent than real. That is, although enclosure in a legal sense had occurred, physical reclamation and 'improvement' of the areas in question may not yet have taken place. Alternatively, as suggested earlier, it is possible that Faden reused for the west of the county, where the majority of examples occur, the unpublished survey made by Mostyn Armstrong in order to save on the costs of surveying (although this can scarcely account for the appearance of commons in places enclosed before the late 1770s, such as Brancaster or Great Bircham). Either way, it is evident that Faden shows us *most* of the commons which had existed in the Middle Ages, although not all of them; and that he shows boundaries which in *most*, but by no means all, cases had remained unchanged since that time. It is thus possible to use Faden's map to tell us a great deal about the character and origins of Norfolk's common land, and also much about the development of settlement in the county in the Middle Ages and after. This is because in Norfolk the history of commons, and settlements, was intimately related: before the eighteenth century a very high proportion of farms and cottages in the county were located on the edges of a common, or on a road leading directly to one.

The origins of this characteristic settlement pattern were first discussed by Peter Wade Martins in the 1970s and 80s, and further explored by Alan Davison, Andrew Rogerson and other field archaeologists over the following decades (Wade Martins 1980; Davison 1990; Rogerson 1995).[7] In Norfolk, as elsewhere in England, early Saxon settlements had generally been small, dispersed and relatively mobile in character, lasting only a generation or so before being relocated elsewhere (Hamerow 1991). But in middle Saxon times – probably from the later seventh or eighth century – larger and more stable settlements developed, which in Norfolk are marked in the ploughsoil by scatters of a distinctive pottery called Ipswich Ware. Most of these are found close to parish churches. These sites continued to be occupied into late Saxon times, most of the churches themselves probably coming into existence then. But settlement often then drifted away to the edges of greens and commons, in some cases leaving churches quite isolated in the landscape, alone in the fields. Wade Martins originally suggested that this was essentially a post-Conquest phenomenon – commons 'were not the focal points of settlement before the twelfth century' (Wade Martins 1980). But more recent work, most notably by Rogerson in Fransham, leaves little doubt that migration had, in fact, begun well back in the eleventh century, although it certainly continued, and in some cases intensified, through the twelfth century and into the thirteenth (Rogerson 1995). As we shall see, close examination of Faden's map can amplify and to some extent qualify this well-researched picture.

The distribution and morphology of commons

Faden's map does not quite show us the total distribution of common land in the county in the 1790s, for one or two commons were apparently omitted (such as that at Feltwell) and a number were included within his category of 'fens and marshes'. Nevertheless, the map does show the size and shape of the vast majority, something which could otherwise only be reconstructed – and then only imperfectly – by the systematic analysis of large numbers of individual enclosure and estate maps – where these survive. No less than 64,756 hectares (160,012 acres) of such 'waste' are recorded on the map, a figure not unadjacent to the 144,846 acres which Nathaniel Kent estimated remained in the county in 1794 (his figure comprised 1,500 acres of 'sedgy and swampy grounds', 80,000 acres of unimproved commons, and 63,346 acres of 'warrens and sheep walks').[8] Faden describes 38,794 hectares of this land as 'common', 16,620 as 'heath', 6,042 as 'warren', 2,519 as 'green' and 781 as 'moor' (Figure 35). As well as cumulatively covering the greatest extent of ground, 'commons' are also the most individually numerous, with 465 areas so labelled: 'greens' come second, with 128 examples. The map shows how the commons of one parish often formed part of long, interconnected chains, which rambled for many miles across the countryside. It would, for example, have been possible to walk all the way from Shipdham to East Harling – a distance of over 21km – without leaving one. Commons usually lay towards the edge of parishes. Indeed, the majority were touched at some point by a parish boundary. The larger commons could be shared between several neighbouring parishes, sometimes without physical division. Mousehold, for example, was exploited by eleven; the great Stock Heath, near Swanton Novers, by ten.

Commons came in a wide variety of shapes and sizes but characteristically exhibited curving, irregular edges which often narrowed towards one or more 'funnels', where roads entered them. Some examples followed water courses – that is, they occupied ribbons of lowlying, waterlogged ground beside streams and rivers. Others were found on level interfluves. On the claylands in the centre and south of the county this was often where there are slight concavities in the plateau surface, giving rise to areas of seasonally waterlogged ground. Elsewhere, commons seem to correspond to patches of particularly infertile or intractable soils associated with glacial drift. Nevertheless, while in *general* terms commons occupied the poorest or most difficult soils in any parish, their distribution and location were not simply 'determined' by environmental factors. Their edges do not correspond in any simple way with changes in soil type, drainage, or topography, and their essentially human, artificial character becomes evident when we superimpose on Faden's map the parish boundaries recorded on the tithe award maps of *c*.1840 (Figure 36). While the latter often run through the middle of an area of common, just as frequently a common will run up to a boundary on one side only: or, to put it another way, the common edge fits precisely within the parish boundary. Evidently, the community living on the other side of the line found it perfectly easy to cultivate the soils in this location, farming the land right up to their side of the boundary and preferring to maintain an area of common grazing elsewhere. In detail then, the location and configuration of commons were the result of human decisions, not the consequence of the unanswerable dictates of topography. This is an important observation, and one which it is only really possible to make because parish boundaries can be superimposed so easily upon the digitally redrawn map.

Faden's map reveals marked variations in the density and morphology of Norfolk's commons. In the south-east of the county, in particular, a high proportion took the form of narrrow strips – little more than widened roads.

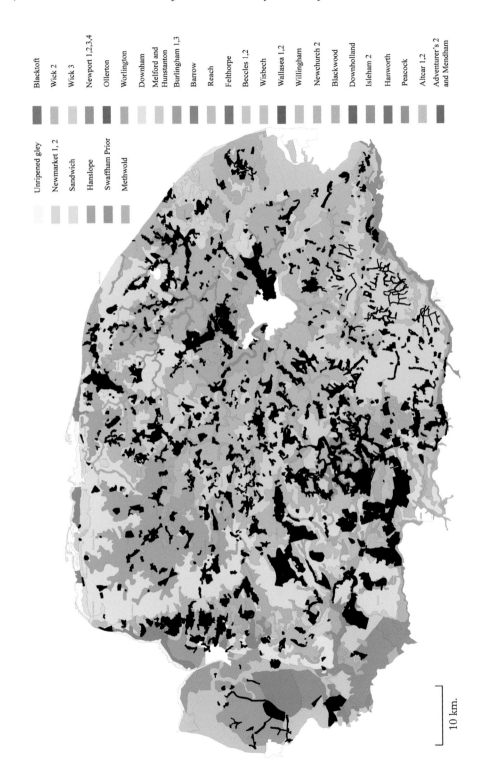

Unripened gley

Newmarket 1, 2

Sandwich

Hanslope

Swaffham Prior

Methwold

Blacktoft

Wick 2

Wick 3

Newport 1,2,3,4

Ollerton

Worlington

Downham

Melford and Hunstanton

Burlingham 1,3

Barrow

Reach

Felthorpe

Beccles 1,2

Wisbech

Wallasea 1,2

Willingham

Newchurch 2

Blackwood

Downholland

Isleham 2

Hanworth

Peacock

Altcar 1,2

Adventurer's 2 and Mendham

10 km.

FIGURE 35. 'Commons and heaths' shown by Faden, and the principal soil types mapped by the Soil Survey of England and Wales.

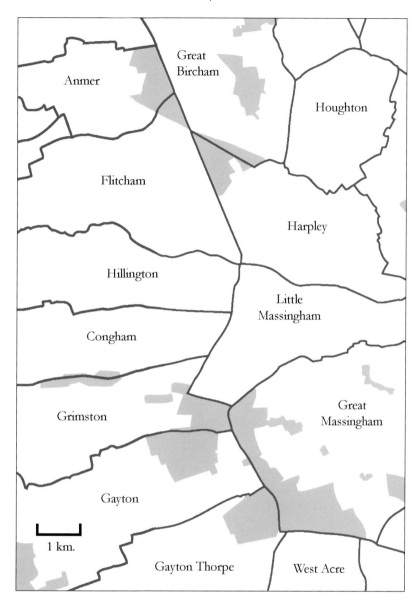

FIGURE 36. The
relationship between
commons and parish
boundaries in the
area around Little
Massingham.

In the south-west of the county, in contrast, in the area between Thetford and Attleborough, commons were more extensive and formed particularly long, interconnected chains, running along valley floors, up onto watersheds, and then down to valley floors again (Figure 37). The largest areas of common land in the county were found here, on the edge of Breckland, and in Breckland itself. Other particularly large and continuous areas existed to the north of Norwich, in the area between Cawston and Stanninghall; and to the east of Norwich, where the great Mousehold Heath still extended as a single block over an area of some 2,500 hectares.

As already noted, Faden uses a number of terms to describe areas of common

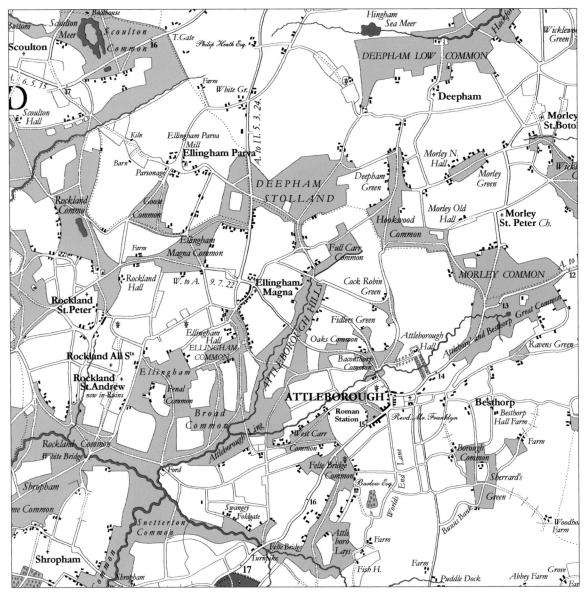

FIGURE 37. Long chains of interconnected commons in the area around Attleborough and Great Ellingham on the eastern edge of Breckland.

land: 'green', 'common', 'moor' and 'heath'. The last of these was a descriptive rather than legal term, and some of the heaths shaded on the map may in fact have been in private ownership and free of common rights at the time the map was surveyed. The same is true of many of the 'warrens', which Faden likewise mainly shows as if they were common. The real significance of Faden's nomenclature is uncertain, however, not least because – as Karen Morley has noted – where it is possible to compare the names which Faden uses for individual commons, with those employed on earlier maps, there are often differences.[9] There is, in particular, a tendency for Faden to describe areas simply as 'Common', prefixed with the name of the parish in which they lie, where earlier maps had used the term 'Green' or 'Moor' and provided more idiosyncratic prefixes. Thus Faden's 'Morley Common' appears on Thomas Waterman's 1629 map of Morley as 'Garsing Common', the former word an old, Middle English term for 'grazing'.[10] Faden's 'Morley Bottom' appears on Waterman's survey as 'Southwoode Moore', while 'Attleborough and Besthorpe Common' is 'Garsing Moor'. In a similar way, the common described as 'Titshall Snath' on a 1727 survey of Tibbenham is named 'Titshall Common' by Faden, while 'The Bushes' has become 'Aslacton Common'.[11] There are many other examples: what a 1590 map of Longham describes as 'South Hall Green', Faden has as 'Longham Green'; while 'Cundalls Green', shown on a survey of New Buckenham dated 1597, is described by Faden simply as 'New Buckenham Common' (NRO MC 22/11). As Karen Morley has pointed out, nearly 87 per cent of commons on Faden's map take their name from a village, but only 59 per cent of 'greens'.[12]

How far these changes towards an essentially simpler nomenclature were part of a more general shift taking place in the course of the eighteenth century, and how far Faden and his surveyors were themselves responsible for changing particular names, is uncertain. It is, however, clear that many areas of common land had more than one name around this time – or, perhaps, names continued to change even in the decades after the map was surveyed. What Faden describes as 'Raven Green' in Wicklewood, for example, Bryant's survey of 1826 terms 'Handkercheif Green'; Faden's 'Bleach Green' in Pulham Market Bryant has as 'Sweeting Green'.

Norfolk is remarkable for the diversity of its soils but commons – and the different kinds of common – were strongly associated with particular varieties. The Soil Survey of Great Britain employs two units for analysing and mapping soils: *series* and *associations*.[13] A series is a soil with a particular mineral content, structure and other characteristics; an association is a group of series which usually occur together. Each association is generally named after the dominant series within it, which in most cases occupies more than 50 per cent of its mapped area. Because soils can vary greatly within a small area – such as a single field – only a few areas of the county have been mapped in terms of series. The following discussion therefore employs the published soil data mapped at a scale of 1:250,000, which is complete for the whole of the county (indeed, for the whole of England).[14] Yet analysing the relationship between features shown on Faden's map, and different kinds of soil, is not entirely straightforward. It is not enough to merely ask the question 'what proportion of commons (or greens, or heaths) fall on this or that soil type', because different soils are not equally represented in the county. If, for the sake of argument, 80 per cent of the county was occupied by one soil type then it would hardly be surprising to find that a very high proportion of commons (or any feature of the landscape) was closely associated with it. In the discussion that follows, therefore, we mainly discuss the location of commons in terms of the *corrected data*, which is an attempt, in effect, to 'fine tune' the raw data by taking into account the frequency or scarcity within the county of the various kinds of soil. For example, those areas

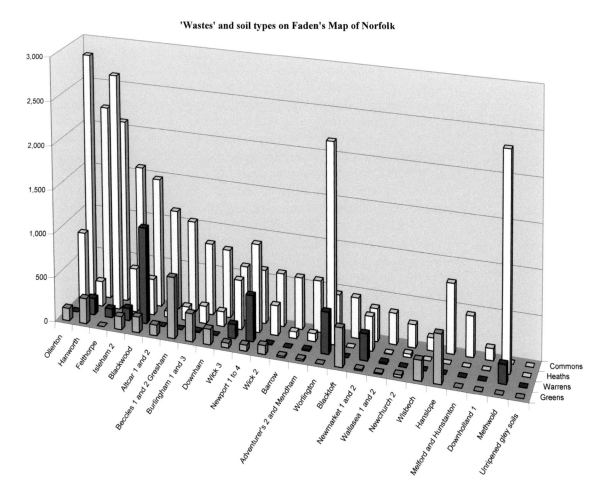

of common land described simply as 'common' display a strong correlation with the heavy, seasonally waterlogged clay soils, or stagnogleys, of the Beccles 1 and Beccles 2 Associations, which characterise the level clay plateaux in the centre and south of the county: with the lighter clays of the Burlingham 1 and Burlingham 3 Associations, which occur on the sides of the main valleys cutting through the plateau; with the stony loams of the Wick 2 and 3 Associations, found in the north-east of the county; and with the dry, chalky Newmarket 1 and 2 Associations, found in the west of the county.[15] When allowance is made for the frequency or scarcity of these soils, however – when we analyse the distribution in terms of the 'corrected data' – the picture changes, albeit only slightly. The association with Beccles and Burlingham soils remains strong, that with Wick 2 and also with Wick 3 reasonably so, but that with the Newmarket soils becomes less marked and that with the sandy, acid soils of the Newport 1, 2, 3 and 4 Associations decreases markedly. 'Commons' were thus mainly found on poorly draining clays, less frequently on stony loams, and only sporadically on light, leached land (Figure 38).

FIGURE 38. The relationship between the different kinds of common land shown on Faden's map, and 'corrected' soil type.

The second largest category of 'wastes' shown by Faden are those described as 'heath'. Indeed, Norfolk probably had – until the later eighteenth century – more heathland than any other county in England. Heath was a descriptive, rather than a legal, term: it signified an area carrying a distinctive type of vegetation, characterised by heather, gorse and other plants which thrive on sandy, acid soils. Heaths were exploited in a variety of way. As well as being grazed by livestock, principally sheep, they were regularly cut for a variety of products. Heather (locally 'ling') was used as fuel, and perhaps as thatch; bracken was cut for bedding, fuel and thatch; while gorse ('furze') was used for both fencing and fuel, especially for ovens.[16] In addition, 'flag' – turf and matted roots of heather – was cut from many heaths, and used for firing by the local poor. Most of the heaths shown by Faden, to judge from surviving remnants, were of the type today categorised by the National Vegetation Classification scheme as *H1 – Calluna vulgaris – Festuca ovina* heath, dominated by heather.[17] Some areas of grass heath also occur in the county, usually where only thin and intermittent sandy deposits overlie chalk (mainly NVC U1). In these, as the term implies, grasses like sheep's fescue are dominant and the characteristic heathland undershrubs constitute subsidiary elements of the vegetation.

The largest concentration of heathland shown on Faden's map was in Breckland, in the south-west of the county, where sands, mainly of aeolian origin, were laid down during the Devensian glaciation over boulder clay or chalk.[18] To the north, a second cluster is shown extending northwards along the edge of the Wash from Mintlyn to Snettisham. This was associated with the Greensand and related formations of Cretaceous date. In earlier times there had been extensive areas of heathland – mainly grass heath – on the higher ground to the west of this belt, in the 'Good Sands' region of north-west Norfolk. But most had been reclaimed during the previous century or so, by large estates keen to embrace the principles of the 'agricultural revolution'. The geology of the area was broadly similar to that of Breckland, but here the layers of sandy drift were thinner, and the soils more easily marled and cultivated.[19] A third area of heathland could be found to the north of Norwich, associated with outwash sands and gravels. Lastly, a string of heaths occupied the moraine deposits along the north Norfolk coast, especially in the area between Holt and Cromer. In addition to these main concentrations, Faden shows a scatter of heaths across the boulder clay plateau in the centre of the county. These were associated with small pockets of glacial sands or gravels lying within the matrix of the glacial clays. Not surprisingly, the 'corrected' soil data shows clearly that heaths were most closely associated with soils of the Worlington Association, and of the Newport 1, 2, 3 and 4 Associations, all poor, acidic, leached and sandy soils. Heaths were also common on the soils classified as the Wick 2 and 3 Associations, which mainly comprise stony loams but which also include some poor sands and gravels. More surprisingly, a significant numbers of Faden's heaths were found on the calcareous loams of the Newmarket 1 and 2 Associations (Figure 38). These must represent remnants of grass heath, formed in thin layers of sandy drift: it is noteworthy that all these examples have since been converted to productive agricultural land, rather than remaining as heath or being planted up as woodland, the eventual fate of many of Faden's heaths.

A number of the areas shaded on the map as common land are, as already noted, described as 'warrens' – i.e., as areas enclosed for keeping rabbits. Many of these may, by the time the map was surveyed, no longer have been true common land – that is, manorial lords may have physically enclosed them to keep rabbits in and other livestock out – but the majority had almost certainly originated as such, for the overwhelming majority of large commercial warrens were established on manorial wastes.[20] The rabbit is not indigenous to Britain – or at least, it has not formed part of our native fauna since before the last Ice Age. The Romans had kept rabbits in Britain, probably in small enclosures, but the animal was only effectively re-introduced from mainland Europe after the Norman Conquest.[21] Small warrens soon became a familiar feature of the landscapes around great mansions and were often placed within deer parks, but larger commercial enterprises had a more restricted distribution. They were concentrated on offshore islands, on coastal dunes, and in areas of inland sand, especially in Breckland.[22] Some of the Breckland warrens may already have covered more than 1,000 acres (405 hectares) by the early fourteenth century, but their area seems to have increased significantly during the agricultural depression of the later fourteenth and fifteenth centuries, in part at the expense of arable land.[23] They continued to increase in numbers, and in area, throughout the sixteenth and seventeenth centuries, partly because they were a good way of making money from unenclosed waste, the law holding that a royal grant of 'free warren' (permission to enjoy exclusive hunting rights over a tract of land) permitted manorial lords to establish colonies of rabbits in such locations regardless of the impact that this might have on the grazing available to the commoners. In the words of the lawyer Harting, 'no action will lie against a lord of the manor for keeping coneys on land over which he has a right of warren'.[24]

Faden shows no fewer than 21 examples of warrens which are clearly coloured as common land (although a few others are shown not so coloured – on the coast by Cromer and Winterton, and on the island of sand and shingle to the north of Wells). Just under half were in Breckland. In many cases, the map shows the specialised buildings called 'lodges' which provided accommodation for the warrener and a place to keep carcasses and skins, as well as all the nets, traps, and other equipment required on the warren. Thetford Lodge, which was erected by the Prior of Thetford to serve Westwick Warren in the early fifteenth century and is now in the care of English Heritage, still survives as the roofless shell of a well-built tower house, constructed of local flint with some brick and limestone dressings (Figure 39). The warren lodge at Mildenhall in Suffolk is similar; there are meagre remains of what is probably a similar building at Ickburgh; and Anne Mason has recently discovered, by visiting the locations where Faden shows lodges, that fragments survive, built into later structures, at a number of other places. At Methwold, for example, what appear to be the

FIGURE 39. Thetford
Warren Lodge, erected in
the fifteenth century by
the Prior of Thetford.

remains of an early stone building, incorporated into derelict eighteenth-century
farm buildings, may be the lodge referred to in a document of 1413 (when it
was roofed with 4,200 tiles).[25] At Santon, similarly, part of a medieval lodge
has apparently been reused as a field barn.

Faden's map shows very clearly the way in which warrens tended to cluster
in groups. Particularly noticeable is the marked concentration to the north of
Thetford, where Wretham, Stanford, Sturston, Tottington and Bodney all lay
adjacent, or in close proximity (Figure 40). In part this pattern simply reflects
the fact that warrens were placed on the very worst soils of the district, 'in
the areas of deeper sands marked by a dominance of *Carex arenaria* [sand
sedge]'.[26] But in part it was a consequence of the fact that the damage caused
by escaping rabbits ensured that neighbouring areas could not easily be used
for much else. When in 1782 the future of Stanford Farm, part of the Merton
estate, was being debated, it was stated that because it was bordered by the
warrens of Wretham and Sturston, it was likely that 'the greater part of it will
be made a rabbit farm'.[27]

High grain prices and a fashionable enthusiasm for reclamation saw the
demise of most of the warrens shown by Faden during the early nineteenth
century – once again, the map captures an important feature of the landscape on
the eve of destruction. The great Methwold Warren, for example, was partially
divided and enclosed by 1824; by 1844 White's Directory was able to describe it
as 'the previous warren … all enclosed'.[28] Nevertheless, several examples survived
into the nineteenth century. Stanford Warren, owned by the Merton estate,
was still stocked with 7,200 rabbits in the 1820s, and rabbits were also being
systematically trapped on the adjoining arable and sheepwalks. Nearly 15,000

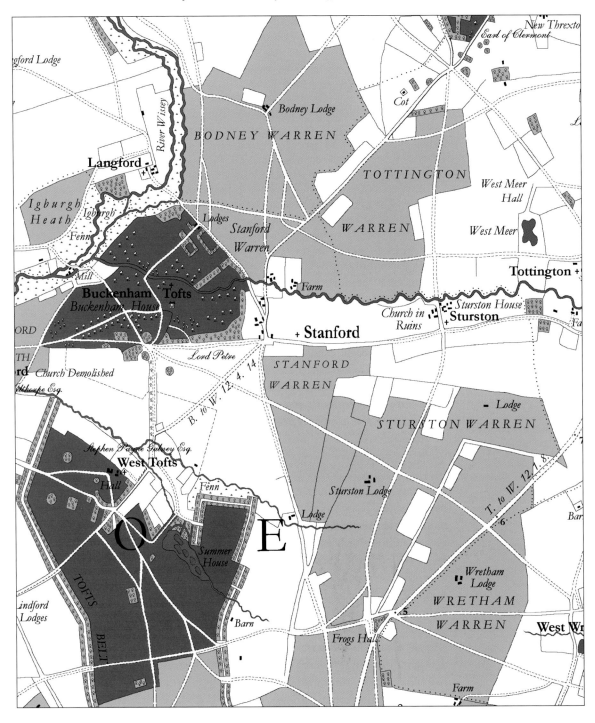

were taken between 17 August 1824 and 1 March 1825.[29] Indeed, one or two of the warrens shown by Faden continued to operate into the early twentieth century, when they were abandoned due to the progressive afforestation of the area by the Forestry Commission.

FIGURE 40. One of several tight clusters of rabbit warrens shown on Faden's map.

When allowance is made for the relative frequency of soils in the county, warrens display a similar pattern of association to heaths, being found in particular on soils of the Newport 1, 2, 3 and 4 Associations; on those of the Worlington Association; and on those of Newmarket 1 and 2 Associations. This is hardly surprising: rabbits flourished on light and especially sandy soils better than elsewhere, not simply because in such conditions they could burrow easily and keep dry, but also because they consumed heather, gorse and other heathland vegetation with relish. One Norfolk rabbit keeper told a House of Commons Select Committee in 1875 that 'Gorse is a very despised article, but I appreciate it very highly'.[30]

Areas described as 'Green' on Faden's map exhibit, as we might expect, a very different pattern of distribution. They have a strong association, in terms of the corrected data, with the heavy clay soils of the Beccles Association, and to a lesser extent with the lighter, sandier clays of the Burlingham Association. The soils of the Hanslope Association, a rather less frequent type of clay soil in the county, also score well in this respect. But while most 'greens' are thus associated with clay soils – and mainly with the heaviest, most poorly draining varieties – a few examples occur elsewhere. In particular, there is a strong correlation with the relatively rare Hanworth Association soils – peaty, silty loams found in river valleys in the north-east of the county – and with the Blacktoft Association, characteristic of the damp silts of the northern Fens, an area settled and farmed (unlike the peat fens to the south) from early medieval times.[31] Both of these soil types are characterised by a high water table, and seasonal waterlogging. 'Green' was evidently a term traditionally given to places with particularly damp soils, and even though many of the areas locally or traditionally described as 'greens' had become 'commons' on Faden's map, this association is still very evident.

The locational characteristics of areas labelled as 'Moor' by Faden are harder to analyse, because only a small number of commons were so described on the map. Most, like Badley Moor to the south of East Dereham, were located on level areas of poorly draining clay soil, mainly of the Beccles Association. Although 'moor' is today usually employed to describe an area of heather-grown upland, the Old English term *mor* had a more general sense of 'extensive tract of desolate ground'.[32]

Careful analysis reveals a number of other interesting differences between the various kinds of 'waste' shown on the map, and in particular between 'greens', 'commons', and 'heaths'. One relates to the size of the areas so described. There are a total of 128 areas described as 'green' on the map, covering in all 2,519 hectares. They range in size from 0.8 to 190 hectares, with an average area of 19.7 hectares. The sixty-two 'heaths', in contrast, cover no less than 16,620 hectares, ranging in size from 9.7 to 2,168 hectares, and with an average size of 268 hectares. The average 'heath', in other words, was over thirteen times the size of the average 'green'. The 465 commons, as we might expect from Faden's rather general usage of the term, fall in between these extremes, with an average size of 83.4 hectares and a range of 0.8 to 1,957 hectares.

More striking, perhaps, are the differences in the character of the settlements found grouped around the margins of these different types of common. Once again the most striking contrast is between 'greens' and 'heaths'. A high proportion of heaths – just under 40 per cent – have no houses at all around their margins, compared with only 4 per cent of greens (areas described as 'common' once again fall in between these extremes, with around 27 per cent lacking any associated dwellings). But in addition, houses were clustered more densely around greens than other kinds of common or, to put it another way, if we were to divide up the surface area of the common between the houses that Faden shows surrounding it, those fronting greens would each receive a much smaller area of land than those fronting on other forms of common, the figures being 2.2 hectares per house for greens, 8.3 for commons, and 31.5 for heaths (the figure for moors is actually higher still, at 89, but the numbers concerned are too small to be statistically significant). In part this pattern reflects the fact that heaths were generally associated with light, porous geological formations, which might in some cases extend well beyond their margins, making it difficult to establish farms and cottages there because of problems of water supply. On the kinds of heavy clay where most greens and the majority of commons could be found, in contrast, water was more easily obtained, from shallow wells. But in part at least it was a function of the fact that the smaller the common shown on the map, in general, the more densely settled its edges: 'greens', as already noted, were on average smaller than the other categories of common land. In spite of the evident tendency for Faden's map to simplify traditional terminology, the term 'green' was evidently used in a particular way, for a relatively small area of common land, with densely settled margins, occupying an area of relatively damp soil.

The morphology of common-edge settlement

As we have already noted, archaeologists have demonstrated how the girdles of settlement around common edges developed in the eleventh and twelfth centuries as houses drifted away from older sites, marked in the landscape today by parish churches. Historians and archaeologists studying the landscapes of Norfolk, and of north Suffolk, have often emphasised the way that churches came to be left completely isolated by this process,[33] but Faden's map, because it shows us so clearly the relationship between churches, commons and settlement across the whole of the county, suggests that this tendency can be exaggerated. While some churches did indeed come to stand at a significant distance away from the new focus of settlement strung around a common edge, in many places they were much less isolated. Many stood on the edge of the village, rather than away from it altogether. They often occupied a site at one end of a settlement, the houses stretching away along a street which gradually widened into a common, an arrangement seen, for example, at Great Ellingham (Figure 37). In such cases, the spread of settlement along the margins of common land seems to have developed as much through the addition of new dwellings, as from the migration of existing ones. In a number cases the church itself stood (and sometimes still stands) on the edge of the common, as at Mulbarton or Hardwick. Such places, as in these cases, often have names suggesting an origin as subsidiary elements in the settlement hierarchy, such as an outlying grazing farm: places which perhaps only became significant concentrations of settlement at a relatively late date, and which thus obtained parish churches only after the development of common-edge settlement had occurred. Such an explanation does not apply in all cases, however. The parish church of Great Massingham, for example – a place whose name almost certainly implies an early importance – still stands beside a very large common. What is particularly striking is that there appears to be an association between the most isolated churches (those located more than 300 m from a settled common edge or other cluster of dwellings) and particular aspects of the natural environment (Figure 41). A high proportion are thus found on the better-drained and relatively fertile loams of the Wick Association (31%) and on the light clays of the Burlingham Association (29%), with far fewer on the heavy plateaux clays of the Beccles Association (15%), only a small number on the freely-draining, acidic soils of the Newport (8%), Worlington (5%) or Barrow (4%) Associations, and few on the light, calcareous soils of the Newmarket 1 and 2 Associations (5%). In addition, as Figure 41 shows, isolated churches often have large areas of waterlogged peat or silt soils lying at no great distance from them. These are intriguing patterns, although ones for which we have, as yet, no convincing explanation.

A number of features of the girdles of settlements associated with commons are worthy of note. One is the way that farms and cottages were not usually distributed evenly around common edges. Instead, they tended to cluster near to an entrance, where a road funnelled in. Fifteen per cent of the buildings which Faden shows (albeit schematically) around commons, 18 per cent beside greens and 29 per cent of those beside heaths lay within 75m of an entrance. More noticeable is the way that the majority of houses tended to be placed on the northern side of a common. In the case of greens, 31.5 per cent of house lay in the northern quadrant, with between 20 and 25 per cent in each of the other three; with commons the figure is 32 per cent; and with heaths, 37 per cent ('moors' again buck the trend, but in numbers too small to be statistically significant). The overall figure, for all areas of common and 'waste', is 32.2 per cent. The main reason for this arrangement seems to be that some commons – those in river valleys, or occupying slight concavities in the surface of the boulder clay plateau – occupy relatively low-lying situations, flanked by rising ground. Medieval farmers chose to place their homes on the warmer, south-facing slopes in preference to those facing in other directions.

One last aspect of commons, and their attendant settlements, is worth noting. In many cases, 'islands' of enclosures and dwellings appear to be cut out of them, sometimes in the centre but often towards their edges, where they are separated from the old enclosed ground by little more than a wide road – an arrangement presumably intended to ensure continuing access to the remaining areas of common. In many places – as around Besthorpe – commons

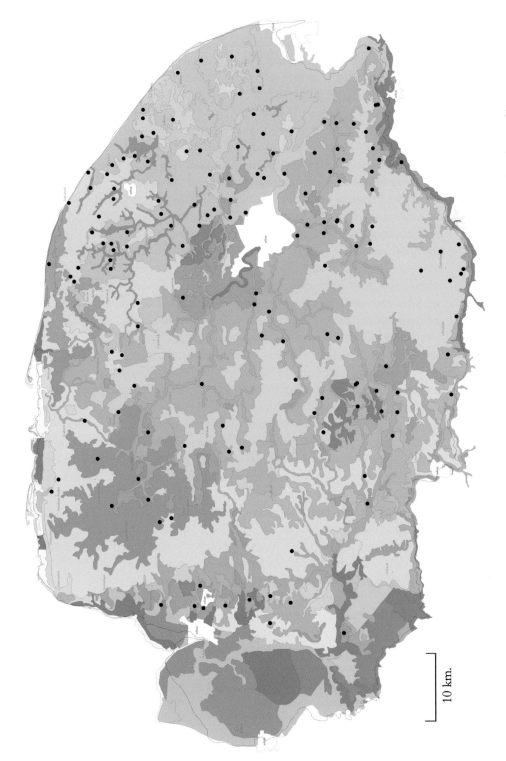

FIGURE 41. Churches shown by Faden as lying more than c.300 metres from a settlement common edge or other cluster of houses.

10 km.

seem peppered with encroachments, numerous 'islands' of private property separated by narrow strips of common land. Some of these, as Addington has noted,[34] appear to have been created in the Middle Ages, but others are later. One of the encroachments on Besthorpe Common is mentioned in an agreement of 1643, by which twenty 'inhabitants of Besthorpe and owners of Estates' permitted John Howse to enclose a portion of the common 'with pale ditche hedge or otherwise'.[35] An apparent encroachment shown by Faden to the north-east of Kimberley Park, sandwiched between 'Downham Common' and 'Wramplingham Common', can be related to an agreement made as late as 1787, by Henry Hobart, 'Lord of the several manors of Wymondham Grishaugh, Wymondham Cromwells and Rustons in Wymondham & the several other Persons owners & occupiers of lands and tenements … to enclose a part of the waste or common of Wymondham containing by measure 11a or 25p'.[36] But not all enclosed 'islands' were cultivated land, or the sites of private houses. The Poor Laws of 1598 and 1601 allowed parishes to enclose portions of common to provide housing or allotments for the poor.[37] The tradition continued into the eighteenth century: Karen Morley has pointed out how examples of Houses of Industry are shown on Faden's map on 'islands' within Summer (Semere) Green in Dickleburgh; on How Common near Gressenhall; on Town Green, Ellingham; and at Watton, Wicklewood and elsewhere.[38]

Notes

1   J. Chapman, 'The extent and nature of parliamentary enclosure', *Agricultural History Review* 35 (1987), pp. 25–35; J. Chapman and S. Seeliger, *Enclosure, Environment and Landscape in Southern England*. Tempus, Stroud, 2001; G. Mingay, *Parliamentary Enclosure in England: Its Causes, Incidence and Impact*. Longman, London, 1997; M. Turner, *English Parliamentary Enclosure: Its Historical Geography and Economic History*. W. Dawson, Folkestone, 1980; T. Williamson, 'Understanding enclosure', *Landscapes* 1, 1 (200), pp. 56–79.

2   J. A. Yelling, *Common Field and Enclosure in England 1450–1850*. Macmillan, London, 1977, pp. 11–29.

3   Turner, *English Parliamentary Enclosure*.

4   M. Turner, 'Parliamentary enclosure', in T. Ashwin and A. Davison (eds), *An Historical Atlas of Norfolk*. Phillimore, Chichester, 2005, pp. 131–2.

5   W. E. Tate and M. Turner, *A Domesday of English Enclosure Acts and Awards*. Reading University Library, Reading, 1978, pp. 178–9, 187.

6   *Ibid.*

7   A. Davison, *The Evolution of Settlement in Three Parishes in South East Norfolk*. East Anglian Archaeology 49, 1990; A. Davison, 'The field archaeology of Bodney, and the Stanta Extension', *Norfolk Archaeology* 42 (1994), pp. 57–79; A. Davison and B. Cushion, 'The archaeology of the Hargham Estate', *Norfolk Archaeology* 53 (1999), pp. 257–74; A. Rogerson, *Fransham: An Archaeological and Historical Study of a Parish on the Norfolk Boulder Clay*. Unpublished PhD thesis, University of East Anglia, 1995; P. Wade-Martins, *Village Sites in the Launditch Hundred*. East Anglian Archaeology 10, 1990.

8   W. Kent, *General View of the Agriculture of Norfolk*. London, 1794.

9   K. E. Morley, *The Origins and Development of Common Land in the Boulder Clay Region of Norfolk*. Unpublished MA dissertation, School of History, University of East Anglia, 2003.

10 NRO, PD3/111.

11 NRO, PET 1009/12.

12 Morley, *Common Land*, pp. 26–32.

13 C. Hodge, R. Burton, W. Corbett, R. Evans and R. Scale, *Soils and their Uses in Eastern England*. Soil Survey of England and Wales, Harpenden, 1984, pp. 52–69.

14 Soil Survey of England and Wales, soil map of England, sheet 4, *Soils in Eastern England*. Harpenden, 1984.

15 Hodge *et al.*, *Soils and their Uses*, pp. 117–22; 132–8; 265–8; 270–9; 346–51.

16 J. Parry, *Heathland*. London, 2003; J. Norden, *The Surveyor's Dialogue*. London, 1618.

17 J. S. Rodwell, *British Plant Communities Volume 2: Mires and Heaths*. Cambridge University Press, Cambridge, 1991, pp. 372–8.

18 G. S. Boulton, F. Cox, J. Hart and M. Thornton, 'The glacial geology of Norfolk', *Bulletin of the Geological Society of Norfolk* 34 (1984), pp. 103–22.

19 S. Wade Martins and T. Williamson, *Roots of Change: Farming and the Landscape in East Anglia 1700–1870*. British Agricultural History Society, Exeter, 1999, pp. 34–43.

20 T. Williamson, *Rabbits, Warrens and Archaeology*. London, 2007, pp. 89–126.

21 J. Sheail, *Rabbits and their History*. London, 1971, pp. 9–10.

22 M. Bailey, 'The rabbit and the medieval East Anglian economy', *Agricultural History Review* 36 (1988), pp. 1–20.

23 K. Sussams, *The Breckland Archaeological Survey*. Ipswich, 1996, p. 96.

24 J. E. Harting, *The Rabbit*. London, 1898, p. 57.

25 J. Sheail and M. Bailey, 'The history of the rabbit in Breckland', in P. Ratcliffe and J. Claridge (eds), *Thetford Forest Park: The Ecology of a Pine Forest*. Edinburgh, 1996, p. 17.

26 G. Crompton and J. Sheail, 'The historical ecology of Lakenheath Warren in Suffolk, England: a case study', *Biological Conservation* 8 (1975), pp. 299–313; p. 303.

27 Wade Martins and Williamson, *Roots of Change*, p. 42.

28 C. Barringer, *Bryant's Map of Norfolk in 1826*. Larks Press, Dereham, 2002.

29 NRO, WLS LXI/23 436 X 6.

30 Sheail, *Rabbits and their History*, p. 50.

31 Hodge *et al.*, *Soils and their Uses*, pp. 124–7; 212–13; R. Silvester, *The Fenland Project, No. 3: Norfolk Survey, Marshland and the Nar Valley*. East Anglian Archaeology 45, 1988.

32 A. H. Smith, *English Place Name Elements*, vol. 2. Cambridge University Press, Cambridge, 1956, pp. 42–3.

33 See, for example, T. Williamson, *Shaping Medieval Landscapes*. Windgather, Macclesfield, 2003, p. 94.

34 S. Addington, 'Landscape and settlements in south Norfolk', *Norfolk Archaeology* 28 (1982), pp. 97–139; p. 129.

35 Morley, *Common Land*, pp. 69–70; NRO, 13964a.

36 NRO, KIM 3/21/5; Morley, *Common Land*, p. 56.

37 S. Birtles, *A Green Space Beyond Self-Interest: The Evolution of Common Land in Norfolk, c.750–2003*. Unpublished PhD thesis, University of East Anglia, 2003; Morley, *Common Land*, pp. 61.

38 Morley, *Common Land*, pp. 61–3.

CHAPTER 5

Woods, Parks and Plantations

Woods and plantations

Woods feature almost as prominently as commons on Faden's map (Figure 42). Areas of woodland in late eighteenth-century Norfolk took a number of different forms, although Faden does not distinguish clearly between them. Some comprised ancient, semi-natural woods, usually dating back at least to the early Middle Ages and which had probably, in most cases, been created by enclosing and managing remnants of the natural woodland vegetation. In the 1790s woods like these were still universally managed on traditional lines, as coppice-with-standards: that is, the majority of trees and bushes within them were cut down to at or near ground level on a rotation of varying lengths, in order to provide a regular crop of 'poles'. The plants then regenerated vigorously from the stump or *stool*, or suckered from the rootstock. The regrowth was vulnerable to browsing livestock and woods like these were invariably bounded by a substantial bank, flanked by a ditch and topped with a hedge or fence (Figure 43).[1] Hazel, so useful for fencing and for the wattle-and daub used in the walls of timber framed buildings; and ash, excellent for tool handles and firewood; were common components of the understorey but a wide range of other plants including lime, hornbeam, elm, maple, hawthorn and holly could also be found, and in particular districts other species might be present, like bird cherry, which is a feature of the ancient woods in parts of west-central Norfolk. The full-grown 'standard' trees, in contrast, were mainly oaks. Traditionally, these would have been widely spaced, so that they did not shade out the coppice growth, and were normally felled at around eighty years of age, when they were of a suitable size for use in the construction of buildings or – more rarely – ships.

The management of woods like this changed over time. In particular, the frequency with which they were coppiced gradually declined. In the thirteenth century most were cut at intervals of six to seven years, but by the nineteenth century this had lengthened to around fourteen. Most of the increase seems to have occurred in the period after *c.*1550 but it has never been fully explained. In part, as Gerry Barnes has argued, it may be due to changes in the demand for fuel. The shift from open hearths to chimneys in the course of the sixteenth century, as open halls went out of fashion and farmhouses became fully floored throughout, increased the demand for larger logs, rather than faggots. But it was

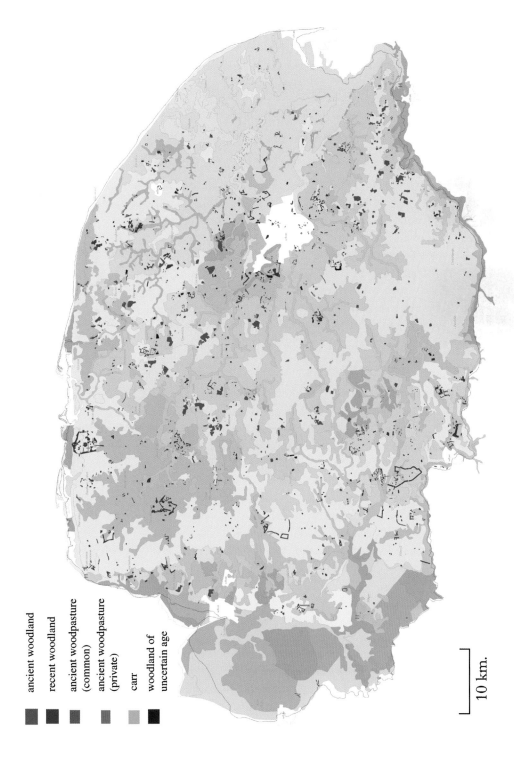

ancient woodland

recent woodland

ancient woodpasture (common)

ancient woodpasture (private)

carr

woodland of uncertain age

10 km.

FIGURE 42. The distribution of different kinds of woodland shown on Faden's map, and the principal soil types mapped by the Soil Survey of England and Wales. (Faden himself does not clearly distinguish between 'recent' and 'ancient' woodland: we have made the distinction ourselves, on the basis of botanical and historical evidence, and locational characteristics).

FIGURE 43. Sexton's Wood, Hedenham, south Norfolk. Part of the wood is being managed on traditional lines, as coppice-with-standards. In Faden's time most ancient woods in Norfolk would have been like this.

also probably because the density of standard trees in woods increased steadily in the course of the post-medieval period, from around 7 to around 20 per acre (*c*.17 to 50 per hectare), thus reducing the vigour of the coppice understorey growing beneath.[2] This was because, with the development of better transport systems through the eighteenth and nineteenth centuries, coal began to replace wood as a major fuel, while iron supplanted it for making many kinds of tools and implements. The value of coppice wood thus declined, but that for timber remained buoyant. The officials who compiled the documents called the Tithe Files in the 1830s and 40s commented on a number of occasions on the poor condition of local coppice, noting at Fulmondestone, for example, that it would have been 'much better if timber was thinned'.[3]

Although the majority of the woodland shown by Faden was of this ancient, semi-natural type, a large amount was more recent in date – it had been planted during the previous century or so by the owners of large landed estates, partly for economic reasons, partly as game cover, and partly to beautify the landscape. Many of these woods lay in or around the landscape parks which formed the settings for the homes of such men. Some were managed as coppices but the majority were plantations, that is, they consisted entirely of timber trees, and lacked any coppiced understorey. Plantations may have existed in early times but, so far as the evidence goes, they were first established on a significant scale in the seventeenth century and then, in increasing numbers, through the eighteenth and nineteenth centuries, often following the enclosure of heaths and other commons. Forestry was a new way of using such marginal land. Faden's map does not distinguish systematically between ancient, semi-natural woods, and post-medieval plantations, but some of the latter are indicated by nomenclature, the map helpfully labelling twenty-two examples as 'plantations' and 'fir plantations' (totalling 177 hectares), four as 'fox covers (9.1 hectares), one as a 'nursery' (3 hectares) and three as 'clumps of firs' (41.2 hectares). Only by comparing the map with other sources, such as English Nature's Register of Ancient Woodland, estate records and other early maps – or by considering the present composition of particular woods, or their location – is it possible to suggest which of the woods Faden depicts fall into which categories.

It is useful to distinguish a third type of woodland shown on the map: 'carr' or wet woodland, dominated by alder and to a lesser extent willow, which occupied, and still occupies, areas of peaty or silty soils on the floodplains of Norfolk's rivers, especially in Broadland. Woodland of this type is relatively easy to identify, partly because Faden often labels it as 'carr' but also because of its location. Around 1,426 hectares of the woodland shown by Faden was of this type, just under 11 per cent of all woodland represented on the map. Although some carr woodland was probably very ancient at the time the map was surveyed, there is good evidence that much was recent, and that in general this kind of woodland was fairly short-lived, developing at the expense of fens or meadows, and lasting for a few generations before turning back to fen or meadow once again, for reasons which remain unclear.

Ancient woodland

As we have already noted (above, p. 86), Faden's map is not very reliable in its depiction of ancient woods, especially in parts of the south-east of the county. Several of the smaller woods are missed out altogether, and the boundaries of some of the larger ones appear inaccurate. Moreover, most of the smaller woods are not individually labelled, and many of the larger ones are named wrongly: two quite different areas are thus described as 'Hedenham Wood' – Hedenham Wood itself, lying to the north of the parish church, and Sexton's Wood, lying some way to the south, which has (so far as we know) never been known by this name. Nevertheless, the map is of some value in showing the overall distribution of ancient woodland at the end of the eighteenth century. This is particularly important because, although Norfolk's ancient woods survived the following century better than its commons, a surprising number of examples were grubbed up, and the land they occupied turned over to agriculture, during the Napoleonic Wars and the Victorian 'high farming' period, especially the smaller 'groves' on the claylands in the south.[4] This destruction reflected both rising grain prices and the fact that with the advent of new agricultural techniques, such as improved methods of field drainage and large-scale marling, the heavy clays and acid sands on which most woods grew could now be cultivated economically. But in addition, in the course of the eighteenth and nineteenth centuries large landowners became increasingly hostile to traditional forms of woodland management, not least because the value of underwood continued to decline. They preferred to establish new plantations, rather than maintain what were increasingly regarded as examples of an archaic form of land use.

Faden's map suggests that ancient woodland still covered around 3,674 hectares in Norfolk in the 1790s – around 0.7 per cent of the county's total area. The overwhelming majority was to be found on the heavy clay soils of the Beccles Association, with rather smaller amounts on the lighter clays of the Burlingham Association, and with a significant quanitiy on the acid sands of the Newport 3 and 4 Associations (Figure 42). This is much the same, in broad terms, as the distribution of such woodland in the county today (Rackham 1986). What Figure 42 also brings out very clearly is the tendency for ancient woods to occur towards the margins of the Beccles soils, on the edges of the level clay plateaux: conversely, few ancient woods are found towards the centre of the main blocks of clayland. The reasons for this pattern, which has been noted elsewhere in the country, are unclear, but may be connected with the difficulties of transporting wood and timber in the winter months, for long distances, along rutted clay roads. Early medieval lords may have retained and managed blocks of woodland in more accessible locations, deflecting clearance and assarting into the interior of the main blocks of clay soil.[5]

Faden's map, in spite of its inaccuracies, can tell us other things about ancient woodland. The shape of the examples he depicts is thus significantly different from that of commons. Ancient woods do not exhibit the same curving edges, narrowing to 'funnels', in part because they do not have public roads running in to them. Nor do they ever form long, interconnected chains, although some noticeable clusters occur, where a number of smaller woods appear to have been left through the fragmentation of a larger, more continuous area. Notable examples include the numerous small woods and groves around Kirby Cane, and Stockton, in the south-east of the county.

Recent woodland

Most of the woodland shown on Faden's map was not ancient and semi-natural in character, but recent, the result of the planting activities of major landowners from the seventeenth century onwards. Most comprised plantations in which deciduous species, particularly oak, sweet chestnut and beech, were mixed with a rather larger number of conifer 'nurses' which were gradually thinned as they matured: in the words of William Kent, they consisted of 'Great bodies of firs, intermixed with a lesser number of forest trees'.[6] Faden's map suggests that around 6,977 hectares of land in the county was occupied by such woodland, which thus constituted around 54 per cent of the woods shown on the map, and *c*.1.3 per cent of the total acreage of the county. It is noteworthy that woods of this kind already occupied more ground in the county than ancient woodland. Given that, so far as we know, relatively little planting of woods had taken place before the late seventeenth century, a considerable amount must have been undertaken by Norfolk landowners in the century or so before the map was made.

As Figure 42 indicates very clearly, recent woodland had a rather different, and to some extent complementary, distribution to that of ancient woodland, occupying the lighter lands to either side of (but particularly to the west of) the clay plateau which occupied the centre of the county. In terms of soils, the majority of such woodland was to be found on the acid sands of the Worlington, Newport 1–4, Methwold, Felthorpe, and Barrow Associations; on the thin chalk soils of the Newmarket 1 and 2 Associations; and to some extent on the stony loams of the Wick 2 Association and the loams of the Melford and Hunstanton Associations.[7] Such woods were much rarer on the clays, especially on the heavier stagnogleys of the Beccles Association, and virtually unknown on alluvial silts or peat.

In part this distribution reflects the fact that there was less reason to establish new woods where old ones already existed in abundance. But it also reflects patterns of land ownership. The majority of large landed estates were located on the lighter and often poorer soils of the county, and it was large owners, rather than small freeholders, who undertook significant schemes of post-medieval afforestation. Indeed, much of the recent woodland shown by Faden was closely associated with elite residences, forming belts and clumps in and around parks. Tree-planting was an important symbol of gentility, for only wealthy landowners possessed sufficient land to create woods on any scale. Some of the more famous examples of eighteenth-century estate forestry spring out from the map, including the great belt, and clumps in and around Holkham Park, most of which were still less than a decade or so old when Faden mapped them; and the huge belt at West Tofts, which provided shelter for both the park and the home farm in the open landscape of Breckland. This had been planted in the 1760s by one William Griffin, a Mundford nurseryman, and when the estate was put on the market in 1780 the sales particulars, as well as emphasising the beauty and shelter that the belt provided, noted also the potential economic value of the timber:

> The number of trees that will remain in the Plantations, after they are thinned so as to leave them at a proper distance, to facilitate their Growth, will be about Six Hundred Thousand: which in the Course of a few years, will at least be worth a shilling a Tree, and consequently amount to Thirty Thousand Pounds.[8]

In 1796 William Kent believed that the timber in the belt – which had now been 'thinned several times' – was already worth £50 per acre, although it had only cost £10 an acre to plant. Kent also observed more generally that while 'gentlemen of fortune' in Norfolk had carried out much tree-planting 'in their parks and grounds', the planting of 'pits, angles, and great screens upon the distant parts of their estates, which I conceive to be the greatest object of improvement, has been but little attended to'.[9] Kent exaggerated slightly. Around 3,250 hectares of recent woodland was located in, or on the edges of, parkland, with a further 813 hectares either placed beside large houses not yet provided with a park, or just outside the boundaries of a park. In all, this amounted to around 60 per cent of the recent woodland shown by Faden, indicating that a substantial minority was, contrary to what Kent implied, located in the more remote parts of estates.

FIGURE 44. Detail of Faden's map, showing Ovington and Saham commons. Several Norfolk commons still carried a significant amount of tree cover when the map was surveyed.

Wood pastures

In addition to areas shown on the map as enclosed and densely packed with trees, in a number of places Faden shows more widely spaced tree symbols, occasionally scattered over enclosed land, more usually across heaths and commons (Figure 44). These represent areas of 'wood pasture', in which wood and timber were produced but livestock were also grazed. As browsing animals would suppress the regrowth of coppices, this type of management was

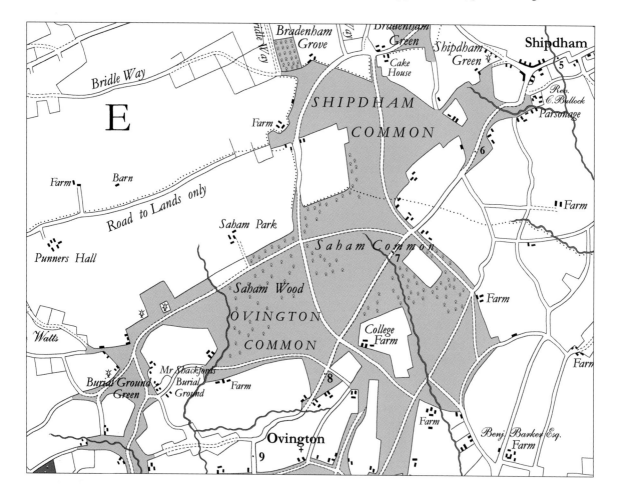

impossible within such areas. Instead, they contained a mixture of standard or timber trees, and *pollards* – in effect, aerial coppices, cut at around 2m above ground level, out of reach of stock. Faden's record of wood pasture is invaluable, because this form of land use was to almost entirely disappear from the county within the next fifty years or so.

There are numerous references to common wood pastures in Norfolk in the Middle Ages and after:what Faden shows us are the surviving fragments of something which was once widespread and familiar. In the thirteenth century, for example, a grant of lands in the parish of Shipdham recorded that the townships of Letton and Reymerston 'had liberty of common for their beasts, with the bishops, horn under horn' in the woods of *Suthaw, Karshaw* and *Blakmere Haw*.[11] It is often assumed that in a relatively densely settled and intensively farmed region like East Anglia wood pastures had generally disappeared by the eighteenth century, especially those on common land. Aside from the fact that older trees might be adversely affected by grazing, or by the compaction of the soil around them by livestock, once trees were felled, died or were brought down by the wind it would be difficult to re-establish them in the face of sustained, intensive grazing pressure – especially where the land was used by large numbers of people who might object to the erection of even temporary enclosures. It is therefore surprising to see so much evidence for this form of land use on Faden's map. Most examples were located on the heavy, poorly draining soils of the Beccles Association, but a significant minority were on the acid, sandy soils of the Newport 3 and 4 Associations, on commons which are explicitly labelled by Faden as 'heath'.

A number of fragments of these wood-pasture commons still survive in Norfolk, in the form of groups of ancient pollarded trees, often preserved within woodland or plantations which were established after the commons in question had been enclosed. Such survivals suggest that Faden only illustrated some, rather than all, such areas. Thus he shows scattered tree symbols across the east and centre of Stock Heath, the extensive common lying between Gunthorpe, Thursford, Barney and Swanton Novers in the north-centre of the county, but none on its northern side. Yet a large number of ancient pollards still exist here within Little Heath Plantation (TF993346), an area of estate woodland established after enclosure on the northern edge of the heath.

The survival of large numbers of pollarded trees on commons and heaths into the later eighteenth century is superficially surprising. But as Patsy Dallas has recently demonstrated,[10] a number of sources show that local communities often had customary arrangements which allowed commoners to plant trees on commons, and to protect them during the early stages of growth. During a legal dispute in the late sixteenth century concerning the commons at Pulham, for example, it was stated that 'the tenantes of the said manor have used to make benefit of the trees growing upon the common near their houses which were planted by themselves and their predecessors'.[12] A manorial survey of 1579, relating to the parish of Gressenhall, similarly records that when tenants were admitted to holdings they received one or more 'plantings', and a detailed map of 1624, by Thomas Waterman, shows these as areas of scattered trees on the commons. Each planting was associated by name with a particular owner and was located close to their home.[13] The practice certainly continued into the eighteenth century. Francis Blomefield, in his *Topographic History of the County of Norfolk* of 1736, described how, in his home parish of Fersfield:

> The tenants have liberty to cut down timber on their copyholds, without licence and also to plant and cut down all manner of wood and timber on all the commons and wastes against their own lands, by the name of an outrun.[14]

Once again, the wooded areas established and exploited by the commoners were explicitly those close to their 'own lands'. Blomefield describes similar customs in the neighbouring south Norfolk parishes of Kenninghall, Diss and Garboldisham.[15]

In this context, it is noteworthy that the scattered trees symbols shown by Faden are often concentrated towards the margins of commons, and that the majority are found close to houses fronting on the common. Fritton Common, for example, has a ring of tree symbols all around its inner margins (Figure 45). Unusually, this common escaped enclosure, and today no less than twenty old pollards can be found here, mostly hidden away among younger scrub and woodland which has regenerated since the intensity of grazing declined in the

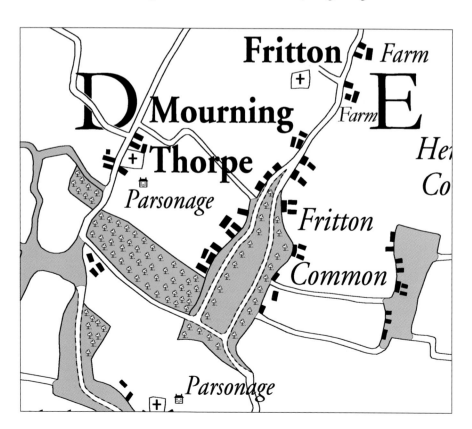

FIGURE 45. Detail showing the way that Faden's 'tree' symbols are concentrated around the margins of Fritton Common.

FIGURE 46. Old oak pollards can still be found in places around the margins of Fritton Common.

course of the last century. All are scattered towards the edges of the common and, for the most part, near to where the houses stand. Most are oaks, with girths in the range 2.3–5.4m, but there are six examples of ash (2.3–3.4m) and, somewhat surprisingly, a sycamore (4.1m). None are particularly old trees and at least half, to judge from their size, must have been planted and first cropped after Faden's map was made. Many were probably cut regularly into the twentieth century (Figure 46).

Fritton is one of only a handful of such survivors. Most common wood pastures were completely destroyed when the commons on which they grew were enclosed, generally (as we have seen) during the two decades after Faden's map was surveyed. Interestingly, the Statement of Claims made in association with particular enclosures often make references to trees growing on commons. That for the mid-Norfolk parish of Shipdham, for example, records that many of the claimants had the right to a 'planting' on the common, on which Faden's map shows a large number of trees. John Platfoot, for example, held three communable messuages and along with rights to common pasture, to cut flags and furze for fuel, and to excavate clay, he also claimed 'the planting of trees standing upon the said common pasture, opposite and adjoining the said premises respectively'.[16] The Earl of Leicester, Lord of the Manor of Shipdham, claimed in respect of five messuages for:

> All trees, and all bushes and thorns planted or set by him or his predecessors, or his or their tenants, upon the said commons and waste grounds, contiguous or near to any of his said messuages or farms, which have been usually lopped, topped, pruned, or cut by him or his predecessors, or his or their tenants.

Some people in the parish made claims for individual trees: John Mendham thus claimed for four trees 'standing and being on the said common, in front of the said messuage'.[17]

Landscape parks

More prominent on Faden's map than any other feature of the landscape, except the various areas of common land, are the parks laid out around country houses, which are themselves indicated by a small 'house' symbol (Figure 47). In the middle and later decades of the eighteenth century, under the influence of designers like Lancelot 'Capability' Brown, Richard Woods and Nathaniel Richmond, it became fashionable to clear away enclosed geometric gardens and walled courts from the immediate vicinity of mansions, and often to remove other formal features in the wider landscape, especially avenues. Great houses now stood, 'free of walls', in open, 'naturalistic' landscapes of grass, irregularly scattered with trees. Sometimes these included a lake of irregular or serpentine form in the middle distance, and clumps of trees, and they were often surrounded in whole of part by a belt of woodland which served to block out the more immediate or less aesthetically pleasing features of the working countryside.[18] A few of these 'landscape parks' developed directly from medieval deer parks, as at Hunstanton and Melton Constable. A rather larger number were adapted (and usually expanded) from deer parks of sixteenth-, seventeenth- or early eighteenth-century date, as at Felbrigg, Blickling, Beeston St Lawrence, Hethel, Buckenham Tofts, Heydon, Merton, Raynham, Stow Bardolph, Ditchingham, Holkham, Houghton, Langley, Kimberley, Wolterton, Earsham, and probably Hanworth and Great Melton. But the overwhelming majority of the parks shown on the map had been created in the period after 1760, usually at the expense of productive farmland.[19]

Parks varied greatly in size, ranging from huge designs like Holkham, extending over more than 1,200 hectares, to tiny examples, little more than paddocks. For a number of purposes it is useful to divide them into three broad size categories: the largest, extending over more than 150 hectares; medium-sized examples, in the 75–150 hectare range; and, most numerous of all, the smaller designs, covering less than 75 hectares.

Faden often seems to have depicted parks with a fair degree of accuracy, especially the larger examples. Indeed, the map can be a very important source of evidence for the history of particular designed landscapes, not least because a large number are not depicted on any other survey earlier than the Ordnance Survey 2" draft of 1813–21. Sometimes, as at Holkham, Faden's surveyors arrived in the middle of a major phase of growth or alteration, which in this case continued right through the 1780s and 90s. This involved a significant expansion of the park, mainly but not exclusively to the south; the establishment around it of a new perimeter belt; and the planting of a number of large new clumps; all of which was carried out under the direction of the head gardener, John

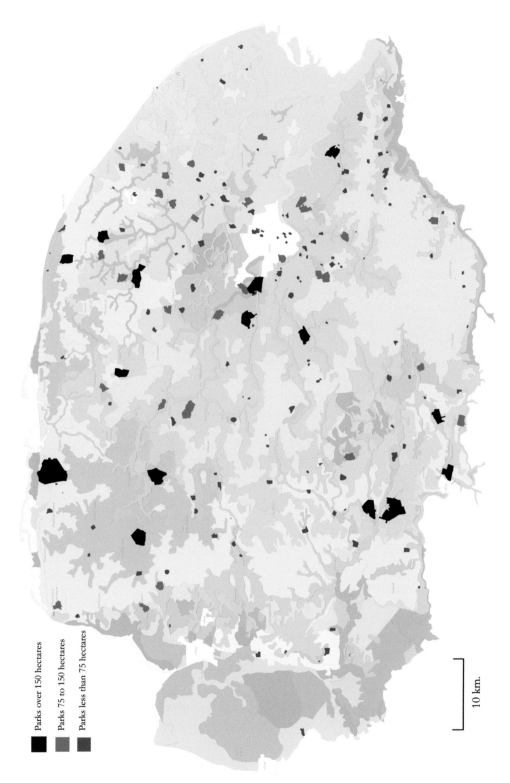

FIGURE 47. The distribution of different sizes of landscape park shown on Faden's map, and the principal soil types mapped by the Soil Survey of England and Wales.

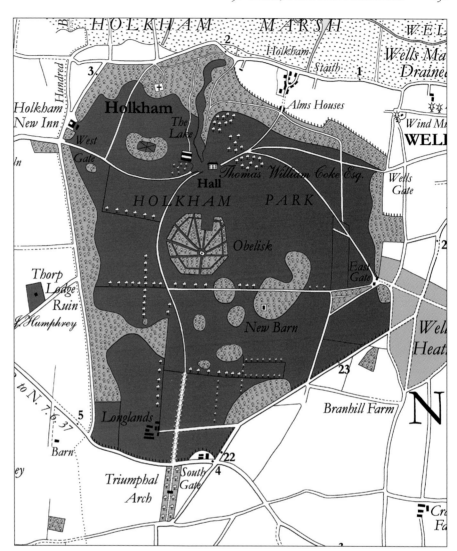

FIGURE 48. Holkham Park, as depicted on Faden's map. Although some of the details are simplified and inaccurate, overall considerable effort has been made to show the layout of this great landscape accurately.

Sandys (Figure 48).[20] Comparing Sandys' planting book, which survives in the archives at Holkham Hall, with Faden's map reveals that the latter shows those clumps and belts which were in place by 1793/4, but not those planted after this date. Comparison with earlier and later maps also shows that the location of the clumps and the configuration of the belts as shown on the map are broadly correct, as is the pattern of drives within the park, although somewhat simplified. The lake is likewise depicted with reasonable accuracy, complete with the marked 'twist' to the north which had been added by the designer William Emes in 1784 (but without the similar twist at the southern end, which was not to be added until 1802). Unusually, lines of trees and hedges are shown within the park, another confirmation of the map's general accuracy: for much of the new southern extension created in the late eighteenth century, although lying within the great perimeter belt, remained under arable cultivation, a

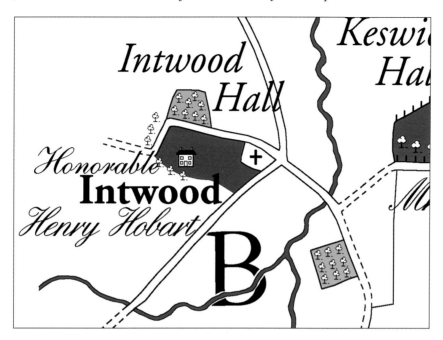

FIGURE 49. Intwood Park, as depicted on Faden's map. Most of the smaller parks are shown schematically, and often inaccurately.

reflection in part of Thomas William Coke's renowned interest in agricultural improvement.

In other cases, however, Faden's map is demonstrably less accurate, and less useful as a source for garden history. The shape and size of some of the smaller parks, in particular, are shown inaccurately, and the details of their layout provided are at best schematic. The map's treatment of Intwood, to the south of Norwich, is typical. Faden shows only the house, the area of the park, the church on its eastern edge, an area of woodland to the north and a lane running along the park's northern margins (Figure 49). The latter, however, is really the entrance drive and should lead to the hall, but this has been wrongly positioned in the middle of the park, rather than on its northern edge. The area of woodland, as we know from a roughly contemporary estate map (Figure 50), is shown much larger than it really was, and the park itself extends too far to the south.[21] Faden's representation of Gawdy Hall, in the south of the county, appears similarly slapdash when compared with maps of the park and associated areas of woodland made in 1789 and 1839.[22] Faden thus shows that the park extended well to the north of the wood called New Grove, but the estate maps show that this in fact formed its northern boundary. To the west, similarly, Faden's park extends beyond the real boundary, marked by Blake's Grove and Lady's Grove. Both of these woods are shown as too large: conversely, what the estate maps depict as Gawdy Hill Wood to the south is shown by Faden as two small woods. The pattern of drives shown by Faden approximates only very roughly to that depicted on the other maps.

Small parks are unquestionably shown less accurately than the larger ones: but even in the case of the latter, the representation is often more of a sketch than a

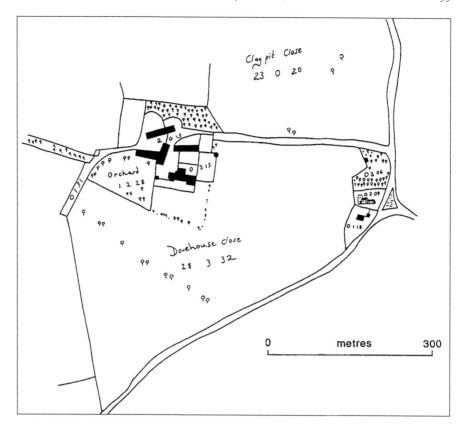

Clay pit Close
23 0 20

Orchard
1 2 28

Dovehouse close
28 3 32

0   metres   300

FIGURE 50. This map, undated but surveyed around the same time as Faden's map, clearly shows the extent to which the latter simplifies and misrepresents the layout of Intwood Park (compare with Figure 49). (Private Collection).

detailed, accurate survey. Houghton Park is a case in point. In 1791 the estate was inherited by an elderly Horace Walpole. He died in 1797 and his successor, the 4th earl, immediately changed the character of the entrance, creating a serpentine drive running in from new lodge gates located at the end of New Houghton village.[23] Other than this, to judge from the estate records, little was done to the landscape between 1790 and 1800 and the representation of the park on a detailed estate survey made by Joseph Hill in the latter year, and that provided by Faden's map, ought to be broadly comparable.[24] In fact, when the two are compared it is apparent that Faden's depiction is fairly schematic, especially regarding the great avenues and vistas created in the early eighteenth century which were (and are) the principal feature of the Houghton landscape. The East View, for example, is shown as a tree-lined avenue by Faden but was in fact a vista cut through woodland; the same is true of the avenue shown running south-eastwards from the house. The layout of avenues to the north of the house is wrong in a number of details, and the overall distribution of woodland within the park is only vaguely correct. Nevertheless, the overall 'feel' of the landscape – still dominated by linear vistas and avenues which would have seemed rather old-fashioned by the 1790s – is correctly conveyed. In general, even when Faden fudges details, or is inaccurate, he at least allows us to see the broad outlines of the design. Moreover, he seldom, if ever, omits prominent features, such as ornamental lakes or even lodges. And once again, what Faden may lack in accuracy, he more than makes up for in completeness, showing as he does every single park that existed in Norfolk in the 1790s. This in turn permits us to analyse the overall distribution of landscape parks in the county at the end of the eighteenth century, and variations in their size and design, in a way that would not otherwise be possible.

Parks were widely scattered across the Norfolk countryside by the 1790s but they were not equally common everywhere (Figure 47). When allowance is made for the relative scarcity or abundance of particular soil types in the county, it is immediately apparent that very little land had been emparked on the extensive areas of peat fen and marsh in the west of the county, none at all on the marshes of Broadland, and comparatively little on the more extensive tracts of level, poorly draining clays of the Beccles Association in the south and centre of the county. Parks were also sparse on the light, chalky soils of the Newmarket Association, on the low escarpment in the north-west. In contrast, much land had been emparked on the light clays of the Burlingham Association and on the sandy, acid soils of the various Newport Associations, and of the Barrow and Worlington Associations. There were also significant areas of parkland on the fertile loams of the Wick 2 Association, and on soils of the Melford and Hunstanton Associations, although the latter is something of a statistical quirk: these soils are very rare in Norfolk generally but Hunstanton and Holkham parks contain a large proportion of their total area. Alluvial river-valley soils, especially those of the Hanworth Association, are also well represented within parks.

The explanation for these patterns is relatively straightforward. Parks, with some notable exceptions, surrounded mansions which lay at the heart of landed estates. Where circumstances discouraged the accumulation of land in large blocks then parks, at least the larger examples, tended to be comparatively rare features of the landscape. Although by the 1790s Norfolk was a largely arable county this was the result of relatively recent, post-1750 changes. In the sixteenth, seventeenth and early eighteenth century large areas of had specialised in cattle husbandry, especially the Fens and the heavy clays in the south of the county. Cattle rearing and dairying were enterprises which tended to favour the survival of the small owner-occupier at a time when, in arable areas, landowners were steadily buying out their poorer neighbours and amassing large, continuous blocks of privately owned land.[25] Looked at in this way, the paucity of parks on the heavy clays of the Beccles Association, and their absence from the various Fen soils, makes immediate sense, although in the latter case other factors were also probably important. The level terrain here lacked the rolling prospects demanded by fashionable taste in landscape, and malaria or 'ague' was still endemic: few gentlemen would wish to make their homes in such a place. To some extent, the fertility and high land values of the Beccles soils also discouraged the build-up of large landed estates, although against this it is noteworthy that a significant proportion of the fertile Wick 2 loams to the north and east of Norwich had been emparked by the 1790s.

What is also significant, however, is that this general pattern changes significantly when we examine parks of different sizes. The smaller parks shown by Faden – those covering less than 75 hectares – show a much closer association with light clays and fertile loams, and are even found to an extent on the heavy soils of the Beccles Association. The large parks, in contrast, are much more strongly associated with dry and acid soils, especially those of the Methwold, Worlington, Barrow and the various Newport Associations, and display an almost total avoidance of the fertile loams and heavy clays, with virtually no examples on soils of the Beccles or Wick 2 Associations, for example. Acre for acre, poor, sandy land was significantly cheaper than rich loams or clays, making it easier to acquire a large block of property. The very largest estates, moreover, were generally found in areas of poorest soils. The medium-sized parks, those covering between 75 and 150 hectares, also avoided the very heaviest clays, and displayed a particularly close association with the poorest of the sandy soils, although they nevertheless also displayed a clear affinity with the fertile Wick 2 loams. The pattern is thus complicated, but there was a clear tendency for the very largest parks to occur on sandy and acid soils, and for smaller parks to display more affinity with heavier and more fertile ground.

In part this complicated picture arises from the fact that soils were not the only, or even the principal, factor structuring the distribution of parklands by the 1790s. Even a cursory perusal of Faden's map shows very clearly that parks were closely clustered in the immediate hinterland of Norwich (Figure 51). This was part of a wider pattern, for similar concentrations of large designed landscapes had, by the later eighteenth century, developed around other major cities in England, and especially around London. Improvements in communications with, in particular, the steady extension of the turnpike system in the second half of the eighteenth century ensured that major towns could be reached with relative ease from their rural hinterlands, which began to change accordingly. People who did not wish to reside within towns were usually keen to live close to them. While long-established landowning families, dwelling in large mansions at the centre of extensive landed estates, generally remained on their ancestral acres the newly rich, or great families whose possessions included land close to a town, often chose to live within easy reach of the attractions and facilities which urban centres had to offer. For not only places like London and Bath, but most major towns, were the centres of the local social universe, and boasted shops, public spaces, assemblies and other opportunities for fashionable consumption and display. Even in the 1720s Daniel Defoe could describe a town like Bury St Edmunds, just across the county boundary, as 'the *Montpelier* of Suffolk … it being thronged with Gentry, People of Fashion, and the most polite conversation'.[26] Of equal importance was the fact that such proximity also allowed those with interests in town – commercial or industrial – to keep a ready eye on business. Indeed, some of the parks shown clustering around Norwich, such as that at Catton owned by banker and businessman Jeremiah Ives, did not have any significant estate associated with them, but simply complimented the appearance of the fashionable 'villas' of the nouveau riches.[27]

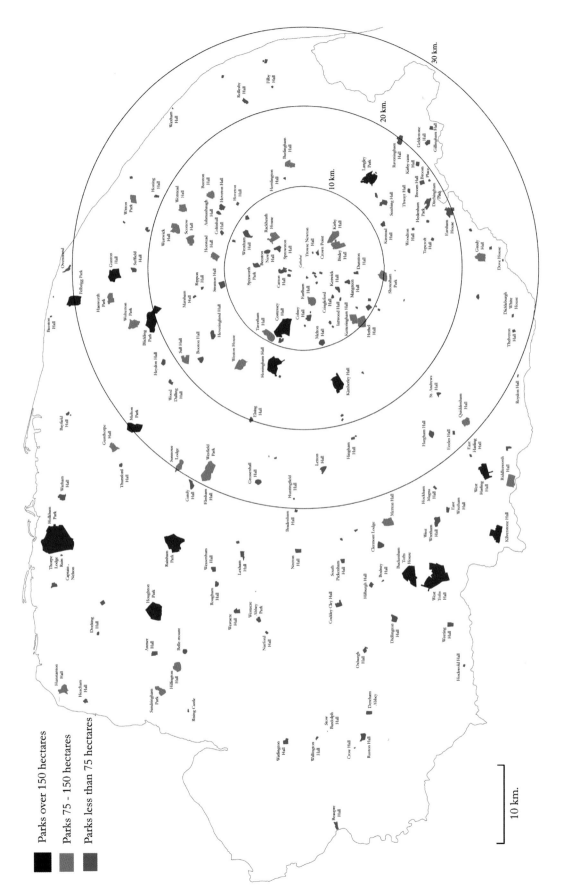

FIGURE 51. The distribution of the different sizes of park shown on Faden's map, and distance from Norwich. The smaller parks cluster noticeably in the vicinity of the city.

Parks over 150 hectares

Parks 75 - 150 hectares

Parks less than 75 hectares

10 km.

30 km.

20 km.

10 km.

No less than 44 per cent of the parks shown by Faden were located within 20km (12½ miles) of Norwich. Not surprisingly, however, parks in different size categories display different patterns in this respect. Whereas 45 per cent of small parks and 53 per cent of medium-sized ones lay within this distance, with the large examples the figure falls to 25 per cent. In part this reflects the fact that land in the immediate vicinity of the county town was, acre for acre, more expensive than that lying at a distance, thus precluding the creation of very extensive parklands. In part, it was a consequence of the fact that the main areas of poor, sandy soil necessary for the creation of the very largest properties lay at some distance from Norwich, principally in Breckland. But it is also because there was some relationship between the size of a park, and its antiquity. The larger parks were, in general, the older ones, associated with long-established families and estates: the smaller parks, in contrast, were more likely to be associated with the nouveau riche, men like Ives.[28] Over 80 per cent of the parks shown on the map in the 150-hectare-plus category had thus been established before 1750: in contrast, only around 23 per cent of medium-sized parks (75–150 hectares) had existed before 1750, and as few as 6 per cent of those covering less than 75 hectares.

There are other noticeable differences between the older and younger, and the larger and the smaller, parks. The older parks, and to a lesser extent the larger ones, were more closely associated with rivers and major valleys than the younger and smaller examples, which had a greater tendency to be scattered across the level interfluves between. This may explain why Faden shows that a higher proportion of the older parks possessed ornamental lakes. No less than 53 per cent of the parks shown by Faden which are known to have been in existence before 1750 had lakes of some kind, as opposed to 8 per cent of those created after this date. But topography – the fact that some parks had significant water courses running through them, which could be dammed to form a lake – is clearly not the only factor in this, for some landowners went to significant lengths to create water bodies even in quite unpropitious circumstances. In the case of Westwick Park, for example, Watts described in 1779 how:

> It was long thought impracticable to obtain an ornamental Piece of Water for the further Improvement of the Scene, on Account of the elevated Situation of the Place, and the nature of the Soil: but the Difficulty is at last fully surmounted; Mr Petre [John Berney Petre II, the owner] having been able, by an ingenious Application of two Archimaedian Screws, to raise a sufficient Supply from a large Reservoir below to the Summit of the Hill.[29]

The screws were powered by a windmill, and a channel three miles long delivered the water to the lake. Wealthy landowners could thus, to some extent at least, override the dictates of the topography while, conversely, even in more suitable terrain lakes were expensive things to construct, involving substantial dams and, in many circumstances, the 'puddling' of the ground to retain water.

Moreover, this difference between younger and older parks in the provision of lakes was not due to the fact that these had been more fashionable in the

period before the advent of the 'landscape' style under Brown and his 'imitators'
in the 1750s. Indeed, the converse was true. We know from other evidence
that while some of the water bodies in question had been created before
1760, as part of geometric or semi-geometric landscaping schemes (Holkham,
1725–31; Raynham, around 1724; Kimberley, expanded to its present size in
1739; Wolterton, 1720s; and Gunton, 1740s), many others were added during
the 1760s or 70s (Melton Constable and Langley in the mid 1760s, both
by Capability Brown; Blickling in the early 1760s; Beeston in the 1770s, by
Nathaniel Richmond).[30] The connection between lakes, and the antiquity of the
parks within which they lay, was rather that long-established parks like these
were owned for the most part by major county families, the kinds of people
who could afford all the fashionable trappings of the 'ideal' landscape. Later
parks, in contrast, were in general the properties of more humble folk. Indeed,
the association between the *size* of parks, and the presence or absence of major
water features, is more striking in this respect. No less than 69 per cent of the
largest parks shown by Faden – those extending over more than 150 hectares
– had a lake; with medium-sized parks (75–150 hectares) the figure falls to 33
per cent; and with parks of less than 75 hectares, it is only 4 per cent. True, to
some extent this last figure may be illusory, in that some of the smallest parks
may have had very diminutive 'lakes', too small to be shown by Faden, given
the scale of the map. This is particularly true where a water feature consisted of
little more than a widening of a stream or river, of the kind sometimes referred
to as a 'broad water' in contemporary documents. That at Bayfield, for example,
almost certainly in existence by the 1790s, is not clearly depicted by Faden. But
this said, the overall trend is clear enough.

Faden's depiction of other features of parkland landscapes, especially in
the case of the smaller examples, is generally too schematic to provide much
information about, for example, significant differences in the design of parks
of different sizes or ages. The map does suggest that woodland formed a more
important component of the design of larger parks, than of smaller ones; and
that the former had, in particular, more extensive and continuous perimeter
belts. Various contemporary or near-contemporary maps suggest that this may
well have been true, but given Faden's often cavalier treatment of the smaller
designed landscapes (coupled, once again, with the scale of the map) we should
not push this too far, or attempt to quantify the difference. Of more interest
perhaps is the way that in some places Faden shows the survival of avenues and
other features relating to the 'formal' geometric styles of the seventeenth and
early eighteenth centuries, most notably at Houghton. Such survivals appear to
have been more common in the larger parks (31 per cent) than in the medium-
sized (13 per cent) or smaller designs (8 per cent). Given what has been said of
Faden's somewhat schematic treatment of the latter we should again not make
too much of this. Moreover, many of the smaller residences shown by Faden
had never possessed any such features, as they had only first come into existence
after they had gone out of fashion. It is nevertheless noteworthy that many of

the greatest landed families in the county continued to maintain vestiges of the old, 'formal' style of landscaping right through to the end of the century, against the tide of fashion. Perhaps the retention of some antiquated features, alongside the adoption of elements of more up-to-date styles, served to signal the long-established nature of the major county families.

Mansions and Owners

Faden marks a large number of major houses – halls, 'old halls', parsonages and other residences of the 'polite' – which did not have a park attached, although sometimes an avenue, enclosed area, or small plantation seem to signify more diminutive gardens and grounds. While 193 large houses are shown with some kind of park, a further 480 are depicted without one, a figure which includes fifty which are specifically labelled 'parsonage'. In fact, two different sized 'house' symbols are used to mark the county's more substantial residences, presumably to distinguish the grand from the not-so-grand, although there are signs that these were not used consistently (there are, in particular, differences in the extent to which the two symbols are employed in different parts of the county). Some large houses were certainly left off the map – at least twenty known examples, including Denver Hall, Sedgeford Hall, and Threxton House. Some at least of these omissions are simply mistakes made when the map was drafted. Letton Hall, built 1783–9, is thus left off the map although its name (and that of its owner) is given, and the park laid out around it clearly shown. But on the whole Faden and his surveyors seem to have recorded the homes of purchasers and potential purchasers with care, except in major towns where the scale of the map generally precluded it.

Around 45 per cent of the large houses shown on the map – 304 out of a total of 673 – are accompanied by the name of their owner. These are a rich source for studying the social make up of the county in the late eighteenth century, but we must remember that Faden was running a map publishing business: although we do not have direct evidence in the case of Norfolk, other county map publishers made a charge for estate and personal names to be engraved on the map, on top of the subscription fee that had to be paid in advance. At the very least we can assume that the named individuals were subscribers. Although most of the residences shown *without* an owner's name seem to be relatively small properties, there are exceptions, including Costessey Hall, home of the Jerningham family, Beaupre Hall and Burnham Westgate Hall. Possibly the owners of these places were unaware of the map's impending publication, perhaps they had more pressing concerns at the time. Alternatively, they may have made a subscription towards Mostyn Armstrong's abortive project, and did not want to be caught out again.

The names on the map clearly signify ownership, rather than residency. Thomas Coke's name is thus attached not only to his great mansion of Holkham Hall, but also to Beck Hall in Billingford; to the family's ancestral seat at Godwick Hall; to Wighton Hall (recently purchased and soon to be demolished); to an unnamed building, probably Dunton Hall, in the parish of Dunton: as well as to the unusual detached deer park at North Elmham. In all, 268 personal names appear on the map – far more than the 96 shown, for example, on Jefferey and Ainslie's 1770 map of Cumbria. Of these, 219 appear to be associated what we would call 'country houses', although these ranged from great mansions like Houghton or Holkham, to diminutive manorial 'halls' little larger than local farms. The 'house' symbols to which the remaining names are attached cannot easily be identified with residences of this kind and presumably represent important houses on village streets, substantial farms, rectories or vicarages.

As noted, the names recorded on the map in association with major residences are, so far as we know, those belonging to people who were prepared to pay for the privilege, either through subscription or otherwise. Thirteen places are associated with a lord, five with an earl, twenty with a knight or baronet, and one with a duke. As several of these names are attached to more than one property, this slightly exaggerates the number of Norfolk's residents in the highest social strata. The overwhelming majority of named individuals – 183, or 68 per cent – are styled 'Esquire', the normal customary title for a substantial property owner, usually bearing a coat of arms – although by this time evidently a term more widely employed, as an honorary description (for even some businessmen like Jeremiah Ives of Catton were so styled). A single property is associated with a 'Gent' and three with people bearing the title 'Honourable'. Below these members of the aristocracy, and of the real or assumed gentry, there are twenty-five people described simply as 'Mr', two 'Captains' (in the sense of retired military officer – a surprisingly low number), two Doctors, eleven people without any title, and no less than twenty-nine clergymen, although not all of these lived in a vicarage or rectory – they include the Rev. Armine Wodehouse of Hunstanton Hall, the Rev. Daniel Collyer of Wroxham Hall, and the Rev. Bacon Bedingfield of Ditchingham Hall. In a property system dominated by male inheritance there are few women owners – a mere eleven individuals are styled as 'Mrs', three as 'Miss' and one as 'Lady'.

Around two thirds of the properties associated with individuals styled 'Mr' lacked any form of park, compared with only *c*.15 per cent of those associated with 'Esquires'. This, nevertheless, means that a third of the places in the former category *are* associated with a park, although usually one of small size. There was thus not an absolute correspondence between the titles carried by individuals, and the character of the properties with which they were associated. The Duke of Norfolk had no mansion in the county, and his name was simply associated with a farm at Earsham marking the site of the ancient deer park of his ancestors; Thomas Coke, the largest landowner in the county and (as noted) associated with no less than five properties, was only an esquire. This said, there is no doubt that the majority of the people named on the map were significant landowners, and many of the others were members of the clergy.

Individuals who were probably businessmen and professionals (people styled 'Mr', 'Dr', 'Captain' or without any title) formed a small minority (less than 15 per cent) of the named individuals, although as noted some of the 'Esquires' probably belong in this group. Many of the substantial houses shown on the map without an owner's name may have been occupied by men of this class, who also constituted many of the residents of the principal towns where, given the map's scale, it was impossible to mark either important houses *or* the names of their owners.

Notes

1   O. Rackham, 'The ancient woods of Norfolk', *Transactions of the Norfolk and Norwich Naturalists' Society* 27 (1986), pp. 161–77; O. Rackham, *The History of the Countryside*. London, 1986, pp. 98–100.

2   G. Barnes, *Woodlands in Norfolk: A Landscape History*. Unpublished PhD thesis, School of History, University of East Anglia, 2002, pp. 214–15, 226–8, 294–307.

3   TNA: PRO, IR 29/5937.

4   Barnes, *Woodlands in Norfolk*, pp. 306–8.

5   A similar pattern has been noted in the Weald of Kent: see K. Witney, 'The woodland economy of Kent, 1066–1348', *Agricultural History Review* 38 (1990), pp. 20–39.

6   Kent, *General View*, p. 87.

7   Hodge *et al.*, *Soils and their Uses*, pp. 107–11; 192–6; 225–7; 245–7; 249–53; 368–70.

8   Kent, *General View*, p. 92; NRO, MC 77/1/521/7.

9   Kent, *General View*, p. 87.

10 P. Dallas, *Wood Pasture in Norfolk: The Relationship Between Trees, Animal Husbandry and the Changing Agricultural Landscape from the First Millennium AD to the Age of Improvement*. Unpublished MA dissertation, University of East Anglia, 2005. Much of this section is based on Dallas's important work.

11 F. Blomefield, *An Essay Towards a Topographical History of the County of Norfolk*. London, 1805, vol. 10, p. 24.

12 NRO, NAS II/17.

13 NRO, Hayes and Storr No. 72.

14 Blomefield, *Topographical History*, vol. 1, p. 95.

15 *Ibid.*, vol. 1, pp. 220 and 263.

16 NRO, BR90/14/2, 4.

17 *Ibid.*, 17.

18 D. Stroud, *Capability Brown*. Country Life, London, 1965; R. Turner, *Capability Brown and the Eighteenth-Century English Landscape*. Weidenfeld and Nicolson, London, 1985; D. Jacques, *Georgian Gardens: The Reign of Nature*. Batsford, London, 1983; T. Williamson, *Polite Landscapes: Gardens and Society in Eighteenth-Century England*. Alan Sutton, Stroud, 1995; T. Williamson, *The Archaeology of the Landscape Park: Garden Design in Norfolk, England, 168–1840*. British Archaeological Reports, Oxford, 1998.

19 Williamson, *Landscape Park*.

20 *Ibid.*, pp. 245–7.

21 Private collection.

22 NRO, 4567 cab II; NRO 222.

23 Williamson, *Landscape Park*, p. 250.

24 D. Yaxley (ed.), *A Survey of the Houghton Estate by Joseph Hill, 1800*. Norfolk Records Society 30, 1988.

25 Wade Martins and Williamson, *Roots of Change*, pp. 21–5, 79–80.

26 D. Defoe, *A Tour Through the Whole Island of Great Britain*. London, 1724, vol. 1, p. 49.

27 Williamson, *Landscape Park*, pp. 225–6.

28 *Ibid., passim.*

29 W. Watts, *The Seats of the Nobility and Gentry in a Collection of the Most Interesting and Picturesque Views*. London, 1779, repr. Garland, New York, 1982.

30 Williamson, *Landscape Park*, pp. 220–3; 237–8; 245–7; 256–8; 259–60; 261–3; 269–71; 286–90.

CHAPTER 6

Fields, Farms and Fens

....

The spirit of improvement

At the time that Faden's map was surveyed Norfolk was famed for its advanced agriculture. 'Turnip' Townshend was by this time long gone, but the activities of Thomas William Coke – 'Coke of Holkham' – were widely reported, and landowners and farmers had embraced a wide range of new agricultural practices which were steadily increasing the productivity of farming, and especially that of cereal growing.[1] The spirit of 'improvement' was in the air: the world could be made a better place through the application of effort, science, and capital.[2] Rural landowners and gentleman farmers were among the map's principal consumers, and Faden clearly recognised that they would wish any map to present the potential for further agricultural change, thus as we have seen carefully delimiting the extent of the many commons and wastes still remaining in the county. But the map also displays, more subtly, some of the achievements that had already been made in modernising the agricultural landscape, as well as the challenges yet to be faced.

'Common fields'

The great wave of parliamentary enclosure in the two decades following the publication of Faden's map not only involved the removal and reclamation of commons and 'wastes'. Arable open fields were also affected, although in many parts of the county these had disappeared well before the 1790s. Some had been removed in the first wave of parliamentary enclosure, in the 1760s, 70s and 80s, which had affected a number of parishes, especially in the west of the county.[3] But many had disappeared long before.

Open fields – areas of arable in which the holdings of farmers were intermixed as narrow, unhedged strips, and over which various shared rights of use and access existed – had dominated the landscape in all parts of the county in the Middle Ages. But these had not, for the most part, been like the highly cooperative and regulated open fields found in the Midlands, in which the lands of all the farmers were thoroughly mingled across the entire area of a parish, and over which strict communal rotations were enforced, so that a third or a half of the land, in a continuous 'field', was left as fallow each year. In most parts of Norfolk holdings were clustered in particular parts of a township, and in many districts communal rights and regulations had always been limited, in some cases virtually non-existent.[4] 'Irregular' open fields like these were particular susceptible to various forms of gradual, informal, piecemeal enclosure.

Throughout the fifteenth, sixteenth and seventeenth centuries proprietors on the heavier soils in the centre and south of the county steadily sold, bought and exchanged strips in order to created consolidated blocks of land which they could then surround with a hedge and ditch and remove from communal routines.[5] By the middle of the eighteenth century some open fields could still be found in certain places on the claylands, especially in mid Norfolk. But for the most part land, other than the commons, was held 'in severalty' – that is, as hedged blocks of private property. On the poorer, lighter lands in the north and west of the county the process of enclosure was slightly different. Here, piecemeal consolidation and enclosure were more likely to be associated with the systematic 'engrossment' of land, as large proprietors – the owners of landed estates – steadily acquired the property of smaller owners, often in time coming to possess the totality of land in a parish. More importantly, enclosure tended to proceed at a slower rate than on the clays, for it was often a more complicated business on this poorer land, largely because of the particular character of the field systems which were found here. The open fields were usually more extensive than on the clays and, in Breckland especially, the holdings of proprietors were often more extensively intermixed.[6] But more importantly, in many places the institution of the 'fold course' existed. Sheep had long formed a key element of the economy of these districts, valued not only for their meat and wool but also for the dung that they produced. Since earliest times the flocks had been grazed by day on the heaths and fallows, and folded by night on the arable land, providing the steady stream of nutrients which, in the absence of chemical fertilisers, was indispensable for keeping these easily leached lands in cultivation. This kind of 'sheep-and-corn' husbandry was widespread in pre-industrial England but in many parts of East Anglia, and especially in these areas of northern and western Norfolk, it took a distinctive form, to which modern historians have given the name 'fold course system'.[7] While tenants might benefit from the manure dropped by the sheep as they roamed over the fallows, the intensive night-folding was reserved for the manorial lord, and could only be enjoyed by the tenant in return for a payment. The organisation of the flocks was completely under the control of the lord, and all sheep had to be placed in the manorial flock, under the care of a manorial shepherd. The fold course system thus originally began, in the Middle Ages, as a way of ensuring that the arable of the demesne – the lord's 'home farm' – received more than its fair share of manure. But in post-medieval times the institution developed in new ways. It gradually became a way of keeping large commercial flocks, from which the sheep of the tenants were now normally excluded. Manorial lords or flockmasters used the institution as a means of monopolising, not so much the supplies of dung, but access to the grazing offered by the commons, and by the harvest aftermath and the weeds on the fallows.[8]

Manorial lords were often loath to forgo this important privilege, and thus rigorously opposed any spread of piecemeal enclosures within the fields which might deprive the flocks of their sustenance. In addition, in most villages there was more than one manor, and more than one fold course, each usually taking the form of a continuous block of arable land and common: 'Within every Towne and village is most commonly ii or ii manors or more and to every manor a Shepps Course or ffouldcourse belongyng'.[9] All this made it hard for open fields to be removed by gradual, piecemeal methods, and instead various forms of large-scale, formal enclosures were often adopted in the course of the seventeenth and eighteenth centuries. As we have noted, most of the parliamentary enclosure acts passed in the 'first wave' of enclosure, in the period before Faden's map was surveyed, seem to have been in the west of the county, and generally involved the large-scale removal of open fields as well, of course, as tracts of heath and common. For a variety of reasons, then, enclosure of the open fields tended to be a more protracted process on the light lands than it was on the clays.

The broad distinction we have just posited between the light soils in the west of the county, and the heavy soils in the centre and south, has many local exceptions, of course. Moreover, there were important districts in which quite different economic or institutional circumstances encouraged, or discouraged, the enclosure of the open arable. But what is clear is that, for whatever reasons, numerous areas of open field still survived in Norfolk in the mid 1790s, if often in small pockets and patches; and it would have been extremely useful to modern historians if Faden and his surveyors had made a systematic attempt to mark such areas on the map, as they so clearly did the surviving areas of common and 'waste'.

In fact, in a few places Faden did blazon the words 'Common Field' prominently across the landscape, sometimes prefixed with the name of a particular village (Figure 52). What is noteworthy is that examples of this practice are clustered in two restricted areas of the county, while conversely, across large areas these words are absent. No 'commons fields' are thus shown across the whole of north-west Norfolk, and only scattered examples are noted on the claylands (at Talconeston, Forncett, Bintry and Foxley) and in the silt fens (near Upwell). In contrast, the map shows a tight cluster of parishes in southern Breckland in which 'common fields' still survived: Rushford, Brettenham, Kilverstone, East Wretham, Riddlesworth, West Harling, Gasthorpe, Bridgeham, Larling and East Harling – with Boughton as an outlier, on the edge of the Fens to the west. Almost as striking is the rather more diffuse concentration in east Norfolk, at Runham, South Walsham, Wickhampton, Norton Subcourse, Woodbastwick and Ranworth (Figure 53).

As always when interpreting Faden's map, it is important to ascertain how far patterns and distributions reflect, not so much real variations in the physical landscape, but inconsistent recording by the surveyors involved in the map's production. Examination of contemporary maps, and details of enclosure awards, certainly suggests that these two clusters do represent districts in which

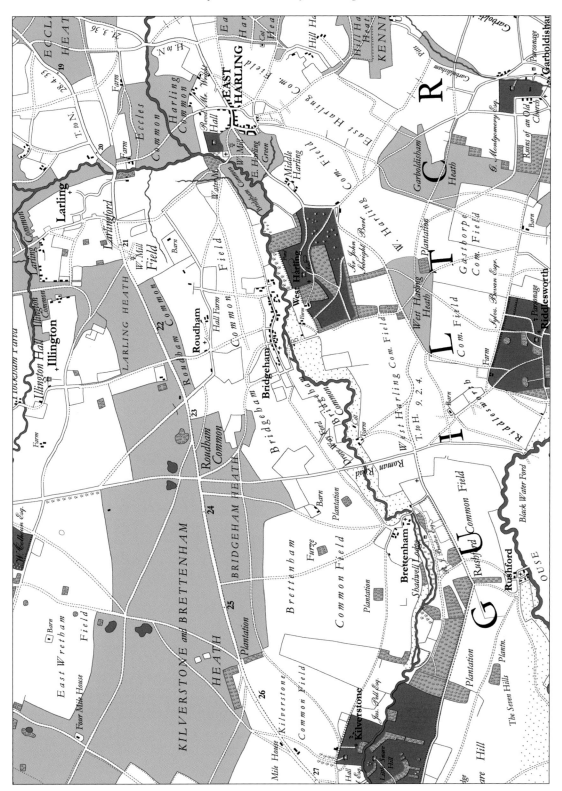

FIGURE 52. A large number of surviving 'common fields' are shown on Faden's map in the area to the east of Thetford.

particularly extensive areas of open fields remained in the 1790s. But such sources also make it clear that the 'common field' label was only applied to *some* of the unenclosed parishes in these two districts. In particular, although the map marks a number of 'common fields' in east Norfolk, many other parishes – especially on the 'island' of Flegg, and on the edge of the Broads – still possessed significant amounts of open arable at this time. We know this because substantial tracts of open field were enclosed here by parliamentary act in the period *after* Faden's map was surveyed. The latter thus makes no reference to open fields in Acle, where extensive areas were enclosed by an act of 1797 (as well as large tracts of common heath and marsh); but it does mark their presence prominently in the neighbouring parish of South Walsham. Only one parish in Flegg has open fields marked on the map – Runham – but the enclosure awards show that extensive areas of unenclosed open arable still survived at this time in many neighbouring parishes. In Repps, for example, more than half the arable still lay open at enclosure in 1805; in Scratby, the majority of the parish still comprised open fields as late as 1842; while in Winterton almost the whole parish consisted of open fields when enclosed in 1828 (there are 295 numbered parcels on the enclosure map, of which 278 are open-field strips).[10]

The appearance of the words 'common field' on the map thus provides a *broad* indication of those parts of the county in which in which open fields survived on a large scale, but does not provide a detailed, nuanced picture – in part perhaps because these words were more easily placed on some parts of the map, less cluttered with other text or information, than others. Why these two areas – east Norfolk and southern Breckland – should have been particularly notable for their open arable is, perhaps, a matter which is beyond the scope of this book, but as so often the answer seems to lie in the character of soils, farming practices and systems of land ownership. East Norfolk was characterised by the particularly fertile and tractable soils of the Wick 2 Association, and from earliest times the way that the local field systems were organised allowed local farmers almost complete freedom in their cropping; communal organisation was often restricted to the right to graze the arable after harvest.[11] There was thus little incentive for the principal landowners – many of whom were Yarmouth businessmen – to undertake the expensive business of enclosure, for it was not hard to find tenants for farms. Even without enclosure, rents in *c.*1800 of between 25s and 27s an acre were normal, rising in extreme cases to 42s.[12] In such circumstances, it is hardly surprising that extensive tracts of open arable still survived. The farming landscape did not, it is true, remain entirely unchanged. Parcels in the open fields gradually increased in size through purchase and exchange, to make more economic units of production. In Repps, for example, the average size of strip more than doubled (from two roods and nineteen perches to five roods and six perches) between 1578 and 1753.[13] But enclosure itself was normally resisted until the high prices years of the Napoleonic Wars made it economically feasible, and sometimes well into the nineteenth century.

In southern Breckland the survival of the open fields seems to have occurred for almost the opposite reasons. Parishes in which 'common fields' are noted were those in which restricted areas of the light, calcareous soils of the Newmarket Association were interspersed with more extensive tracts of poor, acid sands, occupied by heaths and warrens. Most of these places were dominated by large landowners, and had probably escaped earlier enclosure because the low productivity of the land made the expensive business of formal enclosure, by parliamentary act or otherwise, uneconomical, at least until the steep rise in prices of the Napoleonic War years. Most of these places were in fact enclosed without parliamentary acts: landowners eventually bought out any small proprietors that remained, and then reorganised the landscape along more modern lines, often reclaiming the heaths as well as enclosing the arable.[14] Although by the 1830s all common arable appears to have been removed from these parishes, in only one case – that of Bridgeham – does this appear to have been achieved by parliamentary act.

The distribution of open fields in c.1795

The words 'common field' may thus provide a useful guide to those areas of Norfolk in which open fields remained a particularly prominent feature of the landscape in the mid 1790s. But away from these areas of significant survival, as we have already noted, numerous smaller pockets of unenclosed arable land still existed at this time. A long and detailed study of parliamentary enclosure awards dating to the period after the mid 1790s would be able to show where most of these were located, but not all: as the case of the southern Breckland parishes clearly indicates, other forms of enclosure continued to be practised in this period, and these have often left little if any documentary trace. It is possible, however, that a more detailed impression of open-field survival can be obtained from subtle clues on Faden's map. Firstly, of the large number of roads and lanes which it records, a significant proportion are shown as being unbounded by fences or hedges. Some of these were unquestionably private ways running through private fields; others simply ran across unenclosed heaths and commons. But it is likely that most examples were unhedged because they still ran through areas of open-field arable. When the overall distribution of such roadways is mapped, it is noticeable that they are rare in areas in which open fields generally disappeared at an early date, such as the clays of south Norfolk. In contrast, they are clearly concentrated on the light soils of western Norfolk, especially the calcareous loams of the Newmarket 1 and 2 Associations. It was here that areas of intermixed arable had always been most complex and extensive, the fold course system most deeply entrenched, and open fields in consequence most resistant to enclosure, often surviving into the late eighteenth or nineteenth centuries.

Secondly, in a number of places the map appears to show individual field boundaries: never the totality of the field pattern, but only selected elements of it, and often only discontinuous sections of these. Some may represent particularly prominent hedges noted by Faden's surveyors; others may have been included to fill up empty space on the map. But what is striking, once again, is that few examples are shown on the claylands in the south of the county, the area which had long been characterised by a landscape of small enclosed fields. The majority in fact occur in the very same areas in which the 'unbounded roads' are frequent. This suggests that the boundaries may have been included on the map to show where islands of enclosed land existed within an otherwise unenclosed landscape, or where the latter gave way to a private landscape of enclosed fields. Paradoxically, that is, the depiction of enclosed land may indicate the presence of *unenclosed* arable. The distribution of Faden's field boundaries and unbounded roads, together with the words 'common fields', may thus provide a good indication of where areas of open arable still remained in Norfolk in the 1790s (Figure 53).

Michael Turner has mapped the chronological incidence of parliamentary enclosure in the county, and also the distribution of those acts which dealt with some open arable, as well as with areas of common and 'waste'. Figure 54, derived from this work, shows all those parishes in which some open-fields were enclosed by parliamentary act in the period after 1795. The pattern is depicts bears only a very general resemblance to that shown in Figure 53. This, however, is perhaps unsurprising. Turner's published maps make no distinction between places in which only a few acres of residual open field were enclosed by act, and those in which extensive tracts of such land were affected. Moreover, as we have already seen some open fields were enclosed in the period after Faden's map was made by other, non-parliamentary means. On balance, it seems likely that Figure 53 does provide a good guide to the distribution of open arable in the mid 1790s, and that, once again, careful analysis of Faden's map can provide us with important insights into landscape history not easily derived from other sources.

Farms and barns

Apart from fields, buildings labelled 'farm' or 'barn' are the other main features of the contemporary agricultural landscape shown on Faden's map. 'Farms' are widely distributed across Norfolk, but are a particular feature of the claylands of the south, reflecting the highly dispersed settlement pattern characteristic of this area. They were also, however, fairly numerous in the north and west of the county, here reflecting not so much any aspect of the traditional, medieval pattern of settlement but instead the fact that following the enclosure of open fields and commons – a process still, as we have seen, only partially completed in these areas in the 1790s – landowners sometimes built new farmsteads in the midst of their newly enclosed fields, away from the old village nucleations. For obvious reasons, such 'farms' seem, on the whole, to be found in places where enclosure had already occurred before *c*.1795.

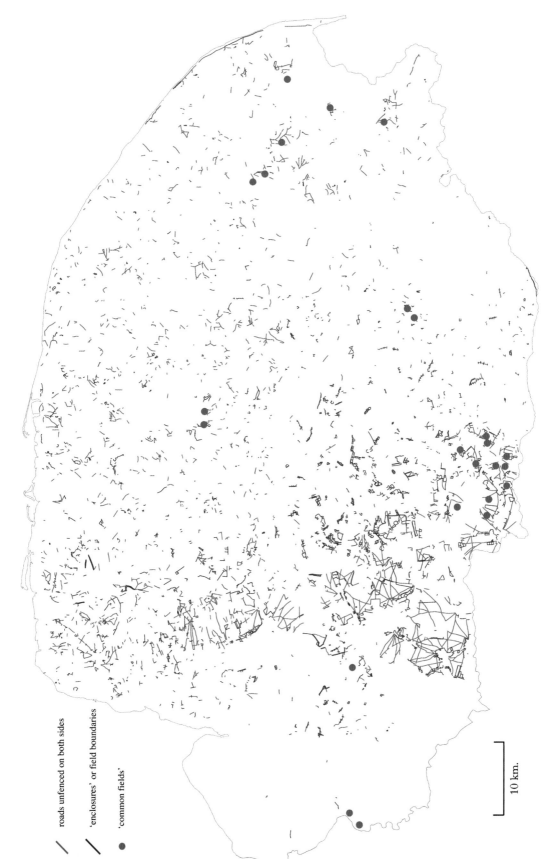

roads unfenced on both sides

'enclosures' or field boundaries

• 'common fields'

FIGURE 53. The distribution of unfenced roads and field boundaries shown on Faden's map, and places where the words 'common field' appear. Together, these features appear to indicate where the principal areas of surviving open-field arable were located at the time the map was surveyed.

10 km.

The isolated 'barns' labelled on the map are a more intriguing feature. They are strongly concentrated in the north and west of the county, again generally in parishes already enclosed (by parliamentary act or otherwise) in the period before 1795 (Figure 55). Faden was clearly using the word to distinguish something other than a normal farmstead, complete with a farmhouse, and precisely what was being indicated is clear from his depiction of Holkham Park. Here, in the area to the south-east of Obelisk Wood, the 'New Barn' is marked. This was the site of a complex of farm buildings designed for Thomas William Coke by the architect Samuel Wyatt and built shortly before Faden's map was surveyed.[15] It comprised a huge barn in Neoclassical style which was originally accompanied by ranges of cattle sheds and other buildings, but which never had a farm house (Figure 56). It was used as a setting for the 'sheep shearings', the agricultural shows which Coke had instituted in the 1780s. While the other 'barns' marked by Faden may have lacked the architectural pretensions of the Holkham complex they were likewise, apparently, groups of farm buildings which lacked an actual farm house.

'Field barns' of this kind were widely used in eighteenth-century England, particularly in areas where extensive heaths and sheepwalks were enclosed.[16] They provided housing for both crops and livestock in circumstances where, as a result of large-scale enclosure, a block of a farm's land lay at an inconvenient distance from existing premises. Cereals harvested from the surrounding fields would be threshed in them, and the straw used for bedding cattle which were

FIGURE 54. The distribution of parishes in Norfolk in which parliamentary enclosures occurring in the period after 1795 dealt with some open-field arable.

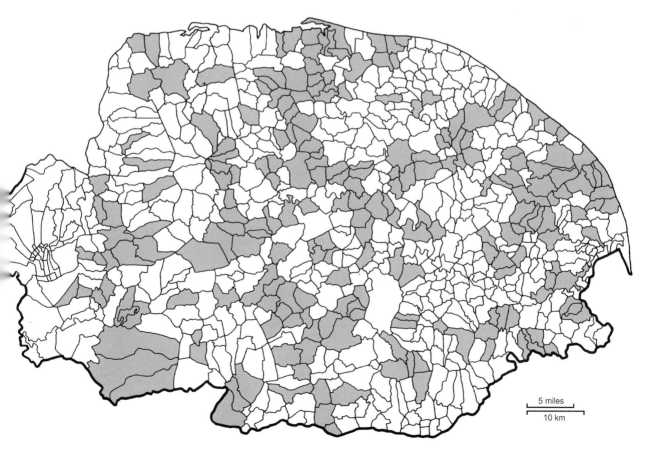

5 miles
10 km

Barns

10 km.

FIGURE 55. The distribution of 'barns' marked on Faden's map.

FIGURE 56. The Great Barn in Holkham Park, designed by the architect Samuel Wyatt and built in the early 1790s, was once accompanied by ranges of cattle sheds. It formed the setting for the 'sheep shearings', the agricultural shows organised by Thomas William Coke (courtesy S.Wade Martins).

kept in attached yards and shelter sheds; the cattle themselves would be fed all winter on the hay and turnips being grown in fields in other phases of rotation. The resultant manure could be spread easily on the surrounding fields. Such an arrangement reduced the distances which crops and manure needed to be carted, and to judge from Faden's map was very widely adopted when the heaths and open fields in the west of the county were enclosed. Such complexes continued to be erected into the nineteenth century, but seem gradually to have fallen from favour during the Victorian 'High Farming' period. Many were then supplied with their own farm houses, or cottages, to allow better supervision of stock.

Wetlands

As well as carefully demarcating the limits of the many tracts of unenclosed common land which still remained in Norfolk, Faden's map also distinguishes numerous areas which are designated on the key as 'Fen and Marsh Lands'. These were concentrated in two main parts of the county: in the east, in the area now usually known as the 'Norfolk Broads', and including the great triangle of the Halvergate Marshes immediately to the west of Great Yarmouth; and in the far west, in the district generally described as 'Fenland' or 'The Fens', a vast level wetland tract which extends into the neighbouring counties of Suffolk, Cambridgeshire, Huntingdonshire and Lincolnshire. In addition, narrow ribbons of such land are shown running along the floors of the valleys of the principal rivers draining the centre and east of the county, the Wensum, Yare, Stiffkey, Tiffey and Bure, and in the lower reaches of their main tributaries; along the north coast from Sheringham westwards to Hunstanton; and down the west coast, beside the Wash. What is striking, however, is that a number of areas which we might well have expected Faden to depict in this way are omitted. Many of the minor areas of poorly drained ground in Norfolk's river valleys are thus ignored. More importantly, while much of Fenland is shown as 'Fen and Marsh', a large proportion in the extreme west of the county, forming a broad band running northwards to the Wash, is not so denoted. This raises important questions about what, precisely, Faden was including in, and excluding from, this category of land.

By lumping 'fens' and 'marshes' together Faden implicitly follows the modern practice of viewing these as much the same thing. But contemporaries were usually keen to emphasise the differences between them. Marsh was the traditional term for areas of coastal silt and clay, usually located within estuaries or behind spits of sand and shingle.[17] In their natural state these were 'salt marshes', flooded at intervals by the sea and with their higher portions carrying a salt-tolerant vegetation featuring such species as sea fern-grass (*Catapodium marinum*) and red fescue (*Festuca rubra*). Unimproved marshes like this provided fairly poor-quality grazing which was traditionally exploited, for the most part, in the summer months, when water levels were lower and the coast less affected by storms. Faden's map shows a number of coastal wetlands of this kind, complete in some cases with the bridges required to move sheep across tidal creeks. But they are interspersed with areas described as 'drained' or 'fresh' marsh, that is, marshes which had been embanked and protected from the ingress of salt water. In a number of cases the map shows the embankments or 'sea banks', as well as the sluices which allowed water to drain out of the marshes when the tide was low. Fresh marshes of this kind were particularly extensive to the west of King's Lynn, where the northern section of what we usually term 'Fenland' or 'The Fens' was traditionally known as 'Marshland'; and immediately to the west of Great Yarmouth – the area we today often describe rather loosely as the 'Halvergate Marshes'. Many drained marshes were used as rich pasture land throughout their history but others were cultivated, especially when cereal prices were high, their fertile soils producing fine crops. Embanking was generally associated with enclosure – that is, drained marshes were usually private property.

Drained marshes have varied and complex histories. Some of those shown by Faden had been reclaimed comparatively recently, in the seventeenth or eighteenth centuries. This was the case with many of the examples along the north coast of the county. But others had been drained for centuries. The district called 'Marshland', for example, was already settled in middle Saxon times, by isolated ranches and salt-making sites, and by the twelfth century a number of villages had developed here, associated with a pattern of small, irregularly shaped fields, protected from the sea by embankments. The wealth and population of the district grew rapidly and by the thirteenth century this was a land of prosperous settlements, with large parish churches. New land was progressively reclaimed inland, onto the lower silt ground, taking the form of bundles of long parallel strips, each no more than 20m wide yet in some cases as much as 2km in length, which ran back towards still unreclaimed areas, used

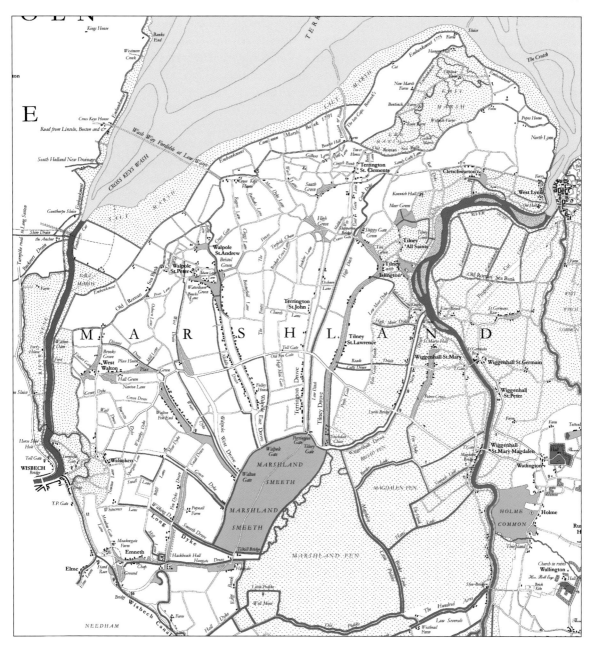

FIGURE 57. 'Marshland' – the siltlands comprising the northern part of the Norfolk Fens – as shown on Faden's map. Note the pattern of long, sub-parallel drove roads, flanked by farms and cottages.

for common grazing, most notably the great Smeeth Fen.[18] These commons were accessed by drove roads which followed the dominant 'grain' of the surrounding landscape, lying parallel with the strip fields. Although Faden's map is not sufficiently detailed to show the fields themselves it does faithfully represent the distinctive pattern of droves that ran through them (Figure 57). This expansion of cultivation in the course of the twelfth and thirteenth centuries was accompanied by the spread of settlement as, in typical East Anglian fashion, cottages and farms appeared around the margins of the droves, and beside other small commons and greens. Again, Faden shows clearly this pattern of settlement, forming long lines running for many miles across the landscape. Much of medieval Marshland was, to judge from the available evidence, under arable cultivation but in post-medieval times the area under pasture appears to have increased significantly. Cattle were fattened here, but also large numbers of sheep, supplying much of the wool for the Norfolk textile industry.

The marshes of Broadland, on the other side of the county, developed in a different way. The main areas of marsh, to the west of Yarmouth, had formed an open estuary in Roman times, the mouth of which was eventually closed by the accumulation of the shingle spit on which the town of Yarmouth now stands. By the late Saxon period the area had become dry enough for small settlements – sheep ranches and salt-making sites – to be established within it, on mounds raised on the low *levees* beside the tidal creeks.[19] Many survive today as isolated farms. Portions of the marsh were steadily embanked in the course of the Middle Ages but this did not become a landscape of arable fields and villages, but continued to be one of isolated grazing farms; and instead of fields comprising long, parallel strips, the old enclosures on the marsh were defined by curving, serpentine dykes. These were the fragmentary remnants of the original pattern of natural creeks and channels of the unreclaimed salt-marsh, which were utilised and thus fossilised as each section of marsh was 'inned' from the tide. Faden's map is drawn at too small a scale to show the details of this intricate landscape, but it does show the scatter of isolated farms – 'Marsh Houses' – which characterised the area (Figure 58). Writing a few years earlier, William Marshall described how:

> The stock are under the care of *marshmen*, who live in cottages scattered over the Marshes: – each having his district, or 'level of marshes', to look after. His perquisite is a shilling upon the pound-rent, which is sometimes paid by the landlord but more generally by the tenant.[20]

The other areas of marsh in the Broads had broadly similar histories but were less extensive – most notably, the tract around the headwaters of the Thurne, in the Waxham–Horsey area, which was (and is) separated from the North Sea by a relatively narrow bank of shingle and sand. The latter features prominently on Faden's map, for although drained and protected from the tides by embankments, marshes remained vulnerable to flooding, and the great storms of 1791 had breached the bank of coastal shingle in a number of places, leading to serious flooding not only in the immediate area but also across the Halvergate Marshes.[21] All this was a matter of great interest at the time, for the marshes remained under threat, and Faden duly shows, and numbers, all nine of the 'sea breaches', and gives details of their individual dimensions on the margins of the map.

So much for the areas of marsh shown on Faden's map. Fens were rather different. They were areas of waterlogged peat soil, rather than clays and silts, and they usually lay inland from the coastal marshes.[22] In the Norfolk Broads, for example, the river valleys draining into the Halvergate Marshes were mainly

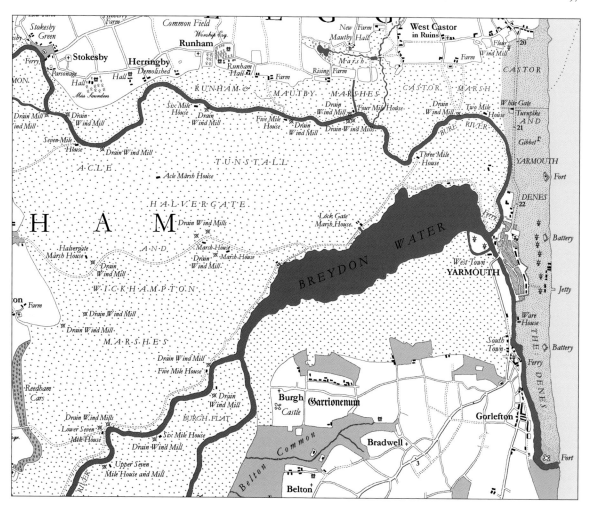

FIGURE 58. Halvergate
Marshes, with scattered
marsh farms and drainage
windmills.

occupied by fens. In the west of the county, similarly, Fenland proper lies inland
from Marshland, and comprises wet peat soils as opposed to estuarine clays.
More importantly, fens were used in different ways to marshes. As William
Marshall explained in 1787:

> The produce and principal use of a fen are totally different from those of a grazing
> marsh. The profits of a fen arise, in general, from Reed and gladdon, cut for thatch
> for buildings. Sedge and rushes, for litter; and thatch, for hay and corn-ricks, and
> sometimes for buildings. Coarse grass, for fodder, and sometimes for pasturage;
> – and Peat for fuel.[23]

In the Middle Ages, and often for long after, fens were usually exploited
as common land by the communities living around their margins, grazed
in the summer months and, as Marshall explains, cut for a wide variety of
products, including peat for fuel. The latter was normally extracted from
shallow cuttings, no more than a spade's depth, but in the Broadland valleys
in medieval times deeper excavation were often made, in order to reach the

more valuable 'brushwood peat', and the resultant pits eventually became the lakes or 'broads' from which the district takes its name.[24] Abandonment and flooding of the pits seems to have occurred in the fourteenth century, in part as a consequence of climatic changes and sea surges, which encouraged flooding, but also perhaps because the collapse of the population in the wake of the Black Death undermined the market for peat, as well as increasing the wages that had to be paid to the workmen extracting it. Once flooded, the margins of the broads rapidly became colonised by reed, saw-sedge, and other emergent vegetation; and as this died, peat began to fill the basins once more, so that broads were often reduced in size with the passing centuries, or even disappeared altogether. Indeed, Faden's map shows a number of examples of such 'lost' broads. Sutton Broad, Carleton Broad, Dilham Broad and Strumpshaw Broad all survived into the early twentieth century, although they are now no more. Others vanished soon after the map was made, such as Honing Broad, Gages Broad, Wiggs Broad, Hare Park Broad and five substantial pools which lay in the area between East Somerton and Horsey. Aerial photographs confirm the accuracy with which some at least of these lost lakes are depicted.

Most of the smaller peatlands in the county, including those in Broadland, remained as undrained commons until the parliamentary enclosure period. But in Fenland itself systematic schemes of drainage were initiated in the seventeenth century. The General Drainage Act of 1600 established the principal that large landowners could overrule local proprietors and any common rights which obstructed the path of drainage projects, and that the investors (or 'adventurers') in such schemes might be rewarded with a share of the reclaimed land.[25] In 1629 the Dutchman Cornelius Vermuyden was employed by a consortium of investors led by Francis, the Fourth Earl of Bedford – who owned considerable properties in the district – to organise a drainage scheme for the southern Fens, work on which began in 1634.[26] Vermuyden, in company with many contemporaries, believed that increasing the velocity of water flowing down the watercourses, by straightening them or amalgamating them or both, would reduce the risk of winter flooding and also ensure that the outfalls to the sea would be kept free of silt. Although a number of improvements to the district's drainage were made, the central feature of Vermuyden's scheme was the construction of the Seventy Foot or Bedford (later the 'Old Bedford') River, which took the waters of the meandering river Great Ouse from Earith, just inside Huntingdonshire, to Denver in Norfolk: a stupendous piece of engineering that runs dead straight for some 32km. Work on all this proceeded rapidly and in 1637, at a Session of the Court of Sewers at St Ives, it was declared that the 'Great Level' had been successfully drained. The declaration was premature. The existing works were insufficient to the task and were frequently vandalised by commoners who objected to the scheme. Charles I himself took over as director of the project and Vermuyden was again employed to remedy the deficiencies.[27] The outbreak of the Civil War suspended all further work until 1649, however, when an act of parliament authorised William, Fifth Earl and First Duke of Bedford and his associates to resume activity, again with

Vermuyden as principal engineer. In this second phase many new drains and embankments were formed but the main achievement was the creation of the Hundred Foot Drain, or New Bedford, River. This ran parallel to the Old River, and substantial banks were created on the outer edges of each in order to create the vast washland which still stores the waters of the Ouse in time of winter flood. The two parallel watercourses mainly lie in Cambridgeshire, but their Norfolk portions are clearly shown on Faden's map (Figure 59). They were as striking a feature of the landscape then as they are today.[28]

In 1653 the Fens were again judged to be drained, and the various investors were allotted their agreed portions of land. Some of the remaining areas of fen common were also divided at this time. But while a proportion of the reclaimed land was now ploughed and sown with crops, there were continuing problems with the drainage works. What the investors and engineers do not appear to have anticipated was that as the peat dried out, it shrank steadily: and where land was cultivated, the soil was constantly being degraded by microbial action. The land surface thus rapidly fell below the level of the adjacent rivers and channels, leading to renewed flooding. In the words of the historian H. C. Darby: 'What seemed a promising enterprise in 1652 had, by 1700, become a tragedy. Disaster abounded everywhere.'[29] The only solution was to use drainage windmills, equipped with scoop wheels, to raise water from the field ditches, over the embankments, and into the main watercourses, and from the 1660s references to such 'engines' appear frequently in the records of the Bedford Level Corporation. Most were simple wooden smock mills with canvas sails, although some smaller structures, resembling the light *wipmolen* of the Netherlands, were also erected: light frameworks of wood, with small sails mounted on an upright hollow post.[30]

As the peat continued to contract, it became difficult to use the reclaimed land as arable, particularly as the late seventeenth and early eighteenth century was a time of stagnant population growth and low cereal prices. By the middle of the eighteenth century most of the peat Fens (and most of Marshland to the north) was under grass.[31] But pasture also needed to be drained, if to a lesser extent than arable, and so the number of mills increased steadily. Some were built by individual landowners, some by the various Drainage Commissions which were established to serve the needs of local landowners in the course of the eighteenth century.[32] During the same period, drainage mills began to be erected in the east of Norfolk, on the Norfolk Broads, again in an attempt to improve the quality of pastures. Initially these were wooden structures but by Faden's time brick mills were becoming common. As early as 1787 William Marshall described the standard Broadland drainage mill as having a body 'built of brick, about twenty feet high, with sails similar to those of a corn mill, but somewhat smaller'.[33]

Drainage windmills were thus a vital technology in the improvement of wetlands and Faden's surveyors appear to have noted the majority of examples then in existence. Indeed, the map is the only source which seems to provide

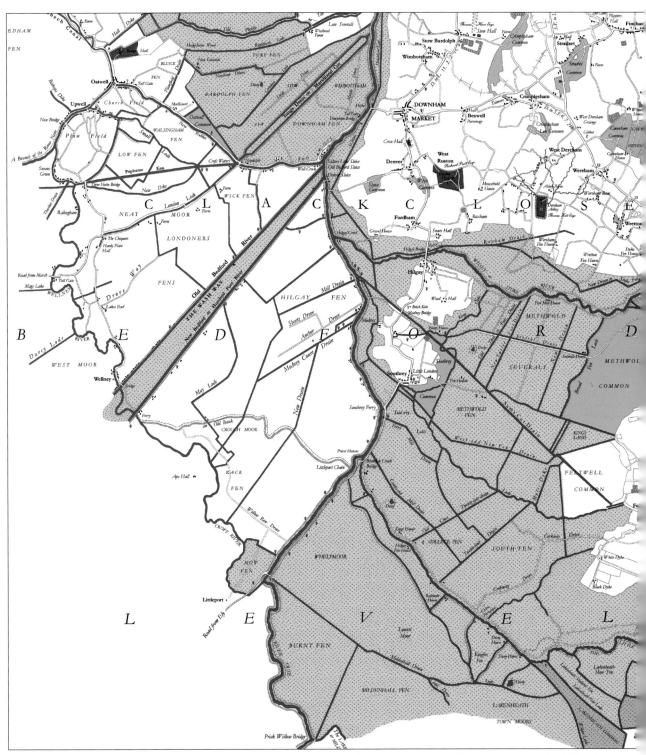

FIGURE 59. The southern fens, showing the distinction made by Faden between 'wetlands' and other areas. Note the Old and New Bedford rivers and the numerous drainage windmills.

a reasonably complete picture of wind drainage in late eighteenth-century Norfolk. Forty-six drainage mills are shown in Fenland, and no less than 47 on Halvergate and in the Broads. Faden actually uses a number of different symbols to represent drainage mills: two specifically for 'drain pump', together with the one employed on the map more generally for corn windmills. In Fenland, the 'windmill' symbol is normally used, with a minority of examples marked using one of the 'drainage mill' symbols. In Broadland, in contrast, the other 'drain mill' symbol is almost exclusively employed, with only a handful of 'windmills'. As so often with Faden, the reason for such variations are obscure. We know that the Fenland mills were usually more massive wooden structures than those in the Broads,[34] and it is possible that the surveyors in this area found it easy to classify them as normal corn grinding mills. A small number of more diminutive structures in the area, resembling perhaps the simple *wipmolen*, seemed clearly different and were accordingly given a different symbol. The Broadland mills were generally lower, less imposing structures than the Fenland mills, but yet were larger than simple machines of *wipmolen* type, and so were accorded a different symbol. The few exceptions in the area which the map marks simply as 'windmill' were perhaps noticeably taller or, perhaps, combined the functions of a drainage engine with that of a grain mill.

No drainage mills of any kind now survive intact in the Norfolk Fenland – they were superseded by steam pumps in the course of the nineteenth century – but they remain a common, indeed iconic, feature of the landscape of the Broads, with more than seventy still surviving in reasonable condition. Very few are the features shown by Faden, although many occupy the same sites. The vast majority were rebuilt as tall brick towers in the course of the nineteenth century, and only a handful survive relatively unmodified: the large drainage mill built within the abbey gatehouse at St Benet's, said to have been erected as early as *c.*1740, and certainly in existence by 1781; the low tower mill at Oby, which carries a date stone of 1753; that on the Brograve level, Hickling, which until relatively recently carried a date stone of 1771; and Highs Mill on the Fleet Dyke in Halvergate, which still contains much of its eighteenth-century machinery. Clippesby Mill and Mautby Marsh Farm Mill may also be the structures which are shown on Faden's map.

While drainage mills were already a common feature of the Broadland landscape by the end of the eighteenth century, they were not uniformly distributed. To judge from the map, largely confirmed by other sources, there were none in the valley of the river Ant, only four in that of the Thurne and none at all in the valley of the Bure upstream of St Benet's. Although it is possible that Faden's surveyors missed some examples this pattern of distribution is probably broadly correct. Most mills – a total of twenty-five – were thus found in the main marshland 'triangle' of Halvergate. Almost all the others were in the lower reaches of the rivers Bure and the Yare. Drainage mills were, that is, largely associated with silt soils, with areas of long-enclosed private grazing marsh, and not with the damp peat commons, although there were one or two possible exceptions. Only in the course of the nineteenth century, as the peat commons were enclosed by a series of parliamentary acts, and often reclaimed, did mills begin to spread higher up the valleys.

At the time that Faden's map was surveyed rising cereal prices had begun to encourage the expansion of cultivation in Fenland. Arthur Young, writing in 1792 about the Suffolk portion of the Fens, described the 'convertible' husbandry currently practised over much of the district. The turf was stripped and burnt, and coleseed and then oats were sown, the land then being laid down to pasture for a few years. He noted that recently, however, a more intensive form of arable farming had developed: 'the common conduct is to make this operation the preparation for successive corn crops, and perhaps in bad rotation'.[35] Although in his account of Norfolk Young implies that the Fens were still mainly under pasture ('one of the richest tracts of grass in Norfolk') Faden's map suggests that much of the land here was, in fact, also now being cultivated.[36] As already noted, while the map shows the eastern side of the Norfolk Fens as 'Fen and Marsh Lands', the western portion is not depicted in this way. This latter area included not only (to the north) the whole of Marshland, but also (to the south) a substantial tract of the peatlands, lying to the south and east of the town of Downham Market (Figure 59). That Faden is indeed depicting a real difference in land use, with much of Fenland now under the plough, is suggested by three things. Firstly, the Fenland peats vary in character and depth, and this potentially arable district generally corresponds with an area in which the thickness of the peat over the underlying clay was, and is, less than elsewhere.[37] Secondly, much of this tract lay to either side of the Old and New Bedford Rivers, a fact which would presumably have facilitated the removal of water from the land. Lastly, although some drainage mills are shown by Faden within the areas denoted as 'Fen and Marsh', they are noticeably more numerous in this area (few mills are shown within Marshland itself, to the north: the silt soils stood slightly higher than the peat, were less prone to compaction on drainage, and could thus usually be drained by gravity, without the use of 'engines').

The fact that Faden suggests very strongly that large parts of Fenland were under cultivation by *c.*1795 may seem a matter of merely antiquarian interest: but, as so often with the map, this is important information. No maps produced before the tithe surveys of *c.*1840 provide us with any real evidence for the pattern of land use in the district, and it is often assumed that the ploughing of the Fens only occurred as steam pumps came into general use in the area in the 1820s and 30s, and as various improvements were made to the arterial drainage system. By showing that conversion of the peat soils to arable was occurring before these developments, the map implies that technological changes were as much a response to, as a cause of, arable intensification.

Indeed, the wetland landscapes of Norfolk, especially the Fens, were constantly changing, as drainage was improved and productivity increased. Faden not only shows us earlier stages of this process, in the form of the various 'old banks' marked in the Fens. He also records intended changes. He thus shows, as we have already seen, the Eau Brink Cut, 3.5km long, which was to speed up the flow of the waters in the lower Ouse by bypassing the 10-km loop in the river above King's Lynn. The scheme had been discussed for many years, in the face of stiff opposition from the population of the town, but an act of parliament had finally been secured in 1795, two years before the publication of the map. Nevertheless, work on the project did not begin until 1817, and was only completed in 1821. The total cost was £500,000, an incredible amount of money.[38]

Notes

1 N. Riches, *The Agricultural Revolution in Norfolk*. University of North Carolina Press, Chapel Hill, 1937; H. E. Saunders, 'Estate management at Raynham, 1661–1706', *Norfolk Archaeology* 19 (1937), pp. 39–76; J. H. Plumb, 'Sir Robert Walpole and Norfolk husbandry', *Economic History Review* 5 (1952/3), pp. 86–9; R. A. C. Parker, *Coke of Norfolk: A Financial and Agricultural Study, 1707–1842*. Oxford University Press, Oxford, 1975; Wade Martins and Williamson, *Roots of Change*.

2 Tarlow, *Archaeology of Improvement*.

3 Turner, 'Parliamentary Enclosure'.

4 B. M. S. Campbell, 'Commonfield origins – the regional dimension', in T. Rowley (ed.), *The Origins of Open-Field Agriculture*. London, 1981, pp. 112–29; B. M. S. Campbell, 'The extent and layout of commonfields in east Norfolk', *Norfolk Archaeology* 28 (1981), pp. 5–32; M. R. Postgate, 'Field systems of East Anglia', in R. A. Baker and A. R. H. Butlin (eds), *Studies of British Field Systems*. Cambridge, 1973, pp. 281–324.

5 Wade Martins and Williamson, *Roots of Change*, pp. 21–5.

6 M. Bailey, *A Marginal Economy? East Anglian Breckland in the Later Middle Ages*. Cambridge University Press, Cambridge, 1989, pp. 44, 50.

7 K. J. Allison, 'The sheep-corn husbandry of Norfolk in the sixteenth and seventeenth centuries', *Agricultural History Review* 5 (1957), pp. 12–30; Bailey, *Marginal Economy*, pp. 65–85.

8 M. Bailey, 'Sand into gold: the evolution of the foldcourse system in west Suffolk, 1200–1600', *Agricultural History Review* 38 (1990), pp. 40–57.

9 Allison, 'Sheep-Corn Husbandry', p. 16.

10 NRO, Csca2/235; K. Bacon, *Enclosure in East Norfolk: Non-Parliamentary Enclosure and Consolidation in the Hundreds of Happing and Flegg*. Unpublished MA dissertation, School of History, University of East Anglia, 1993, pp. 47–61; NRO, C/Sca3/35A.

11 Campbell, 'Extent and layout of commonfields'.

12 A. Young, *General View of the Agriculture of Norfolk*. London, 1804, p. 37.

13 Bacon, *Enclosure in East Norfolk*, p. 61.

14 Wade Martins and Williamson, *Roots of Change*, p. 38.

15 S. Wade Martins, *A Great Estate at Work: The Holkham Estate and its Inhabitants in the Nineteenth Century*. Cambridge University Press, Cambridge, 1980, p. 153.

16 S. Wade Martins, *Historic Farm Buildings*. Batsford, London, 1991, pp. 37, 206.

17 A. Reeves and T. Williamson, 'Marshes', in J. Thirsk (ed.), *Rural England: An Illustrated History of the Landscape*. Oxford University Press, Oxford, 2000, pp. 150–66; C. Taylor, 'Fens', in J. Thirsk (ed.), *Rural England: An Illustrated History of the Landscape*. Oxford University Press, Oxford, 2000, pp. 167–87.

18   Silvester, *The Fenland Project*, pp. 156–69.

19   T. Williamson, *The Norfolk Broads: A Landscape History*. Manchester University Press, Manchester, 1997, pp. 40–8.

20   W. Marshall, *The Rural Economy of Norfolk*. London, 1787, p. 280.

21   NRO, MSC 6/6; NRO, NRS 4193 M9E.

22   C. Taylor, 'Post-medieval drainage of marsh and fen', in H. Cook and T. Williamson (eds), *Water Management in the English Landscape*. Edinburgh University Press, Edinburgh, 1999, pp. 141–56.

23   Marshall, *Rural Economy*, pp. 319–20.

24   J. M. Lambert, J. N. Jennings, C. T. Smith, C. Green and J. N. Hutchinson, *The Origins of the Broads*. Royal Geographical Society, London, 1952.

25   Taylor, 'Post-medieval drainage', p. 147.

26   L. E. Harris, *Vermuyden and the Fens: A Study of Cornelius Vermuyden and the Great Level*. London, 1953, pp. 59–70.

27   Harris, *Vermuyden*, pp. 67–9.

28   N. James, 'The transformation of the Fens', in T. Kirby and S. Oosthuizen (eds), *An Atlas of Cambridgeshire and Huntingdonshire History*. Anglia Polytechnic University, Cambridge, 2000; H. C. Darby, *The Changing Fenland*. Cambridge University Press, Cambridge, 1983, pp. 71–91. C. Taylor, *The Cambridgeshire Landscape*. Hodder and Stoughton, London, 1973, pp. 194–7.

29   Darby, *Changing Fenland*, p. 106.

30   R. L. Hills, *Power from Wind*. Cambridge University Press, Cambridge, 1994, pp. 138–47.

31   Wade Martins and Williamson, *Roots of Change*, p. 30.

32   Taylor, 'Post-medieval drainage', p. 148.

33   Marshall, *Rural Economy*, pp. 282–3. A. Yardy, *The Development of the Broadland Drainage Windmills, with Particular Reference to the Firm of Englands of Ludham*. Unpublished MA dissertation, School of History, University of East Anglia, 2004.

34   R. Wailes, 'The windmills of Cambridgeshire', *Transactions of the Newcomen Society* 17 (1949–51), pp. 97–119.

35   A. Young, *General View of the Agriculture of Suffolk*, London, 1797, pp. 182–3.

36   Young, *Norfolk*, p. 376.

37   Soil Survey of England and Wales, soil map of England, sheet 4, *Soils in Eastern England*. Harpenden, 1984.

38   R. L. Hills, *The Drainage of the Fens*. Landmark Books, Ashbourne, 2003.

CHAPTER 7

For Leisure and Edification

Transport and communication

The kinds of people whom Faden hoped would buy his map were greatly interested in transport, and especially in the state of the county's roads. This was another aspect of the world which, it was widely believed, was in urgent need of 'improvement'. Most roads were in a very poor condition, a matter of concern at a time of economic expansion. But this was a world in which direct state involvement in such matters was inconceivable: only forms of private enterprise could be expected to deliver the changes which most people agreed were required. Until the seventeenth century even the most important roads in England had been maintained by the parishes through which they ran: each parish was solely responsible for its own section, regardless of its population and resources, and regardless of the amount of traffic the road in question had to bear.[1] At the time Faden's map was surveyed the overwhelming majority of roads in Norfolk were still maintained in this way. But from the late seventeenth century 'turnpike trusts' began to be established across England. Created by individual acts of parliament, these comprised boards of trustees which would adopt sections of road, erect toll gates, charge tolls and use the proceeds (after a suitable cut had been taken as profit) to keep the route in adequate repair.[2]

Faden makes a simple division in his key between what he describes as 'Great Roads (Parish or Turnpike)', and 'Cross Roads' (Figure 60). The latter, which comprised the majority of routes shown, were the minor roads and lanes of the county. Most were public vehicular rights of way, although not all, and Faden occasionally informs us that particular examples were private. To the north-east of Ashill, for example, a lane is labelled as 'Road to Lands only': it was probably laid out as an access road to various private allotments at the time of the enclosure here in 1786. The 'great roads' are shown slightly wider than the 'cross roads', and are defined by a heavier line on one side – a practice employed by other publishers of county maps at this time, and one which was later adopted by the Ordnance Survey to show major routes. As we have already observed, where roads of either kind ran through unenclosed land, their margins are shown by dashed lines.

The first turnpike trust in Norfolk was established as early as 1695 to administer the road between Wymondham and Attleborough.[3] But it was well over half a century before others followed. In 1769 the Norwich-Yarmouth

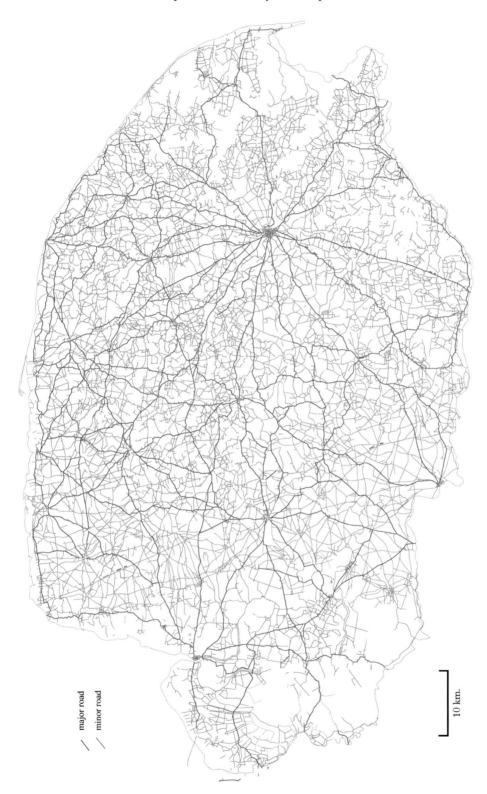

major road

minor road

10 km.

FIGURE 60. The pattern of roads in Norfolk shown on Faden's map.

turnpike was set up to improve and maintain a route running along the line of the modern A47 as far as Acle. It then turned north, passing through Fleggburgh, Filby and Caister, before reaching Yarmouth. The present route from Acle to Yarmouth through the northern part of the Halvergate Marshes, the infamous 'Acle Straight', was only constructed in the early 1830s (this also took traffic from, and thus led to the decline of, the other route from Norwich to Yarmouth, which ran across the Halvergate marshes from Halvergate itself, along the side of the Fleet Dyke, a road which Faden also shows). Also in 1769 the Pye Road – the modern A140 – was turnpiked from Norwich to Coddenham, and in the following year trusts were set up to improve the roads from Norwich to Swaffham; from Norwich to Watton; and from Kings Lynn to Stoke Ferry, Dersingham, and Narborough. The Norwich–New Buckenham turnpike followed in 1772, the road from Thetford to Stoke Ferry was turnpiked in 1791, and in 1794 the turnpike from Norwich to Aylsham was established.[4] Faden's surveyors paid particular attention to these major roads and carefully recorded the gates where the tolls were collected, marked on the map as 'toll gate', 'turn pike gate', or 'TP Gate'. Most of Norfolk's turnpike roads appear to have had two such gates, one near the beginning and one near the end of the route. Thus the Swaffham–Norwich turnpike had one toll gate two miles to the east of Swaffham, and another in the north of Earlham parish, just outside Norwich (the toll house still stands beside the road here, an hexagonal structure, but apparently post-dating the map). The Norwich–New Buckenham turnpike had one gate at Forncett End, four miles west of New Buckenham, and another at Harford Bridge, south of Norwich: the latter also served the turnpike running along the Pye road, the two roads merging (as they do today) just before crossing the river Yare. Turnpike gates often seem, as here, to have been placed near bridges over significant watercourses, in order to reduce the chances of users avoiding the toll (Figure 61). Other examples shown on the map include the gates at Hilgay, Setchey, Wiggenhall St Germans and St Mary, Stoke Ferry, Trowse and Filby.

Not all of the 'great roads' shown by Faden were turnpikes. Indeed, around half were not. These include the coast road running from Hunstanton to Blakeney; the road leading from Docking to Walsingham; and that from Swaffham to Wells. These must still have been maintained by the parishes through which they ran, although we may assume that the county Justices of the Peace, who were empowered to punish local communities that failed to carry out their road duties adequately, paid particular attention to these major arteries of communication. On all the 'great roads', Faden shows at regular points the distance in miles from the start of the road. In the case of turnpikes these often corresponded with the position of milestones, erected by the trust in question, many of which still remain beside major roads (Figure 62). It is almost certain that the numbers on the map do not simply mark the position of milestones, however, as they are also shown on some major roads which had not yet been turnpiked, such as that from Norwich to Fakenham (not turnpiked until 1823)

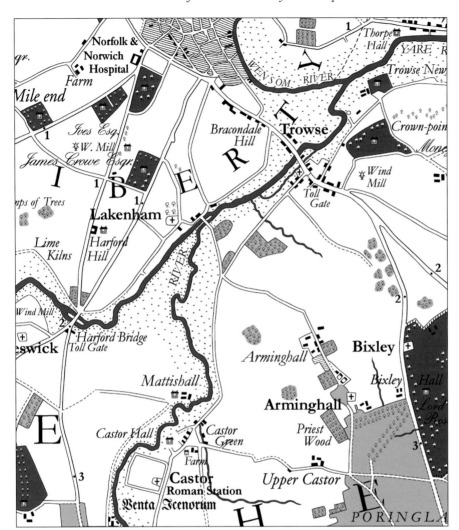

FIGURE 61. The area to the south of Norwich, showing the convergence of the New Buckenham turnpike and the Ipswich turnpike to the south of Harford bridge; toll gates; and mileage figures, probably marking the location of milestones.

or that from Wells to Fakenham (not until 1829).[5] The matter is nevertheless complicated by the fact that some roads which were never turnpiked do seem to have been provided with milestones, presumably erected under the aegis of the Quarter Sessions magistrates. Certainly, Mostyn Armstrong noted in his *History and Antiquities of the County of Norfolk* that 'Besides the turnpike roads, there are some great leading roads which have milestones, and in other respects are no less commodious for travelling'.[6]

Major roads had inns and ale houses at frequent intervals, to cater for the needs of travellers. Other ale houses certainly existed away from the principal transport routes, but the majority of such establishments, and especially those specifically named, are shown beside them. Many still exist, with the same name, such as the Bell at Docking, the Scole Inn at Scole or the Dun Cow in Swainsthorpe. It is evident, however, that the map's coverage of inns and ale houses is distinctly uneven, with the overwhelming majority being shown in the south-eastern

FIGURE 62. Milestone, probably erected in the mid-eighteenth century, beside the old A11 (the turnpike road to Thetford) near Wymondham.

quadrant of the county (a total of 28) compared with only a single example in the south-western, clearly reflecting the interests of particular surveyors rather than any real variations in density. Of those in the south-east, all but six (79 per cent) are on main roads. Other features associated with the county's major roads are also occasionally shown on the map, such as direction posts: we have no idea what proportion of the total existing at the time were recorded.

The 'great roads' connected the principal market towns of the county, which are shown in some detail on the map. Although Norfolk was a predominantly rural county, towns were of vital importance in the economy, not only as manufacturing and service centres but also – in many cases – for their social and administrative roles. All are clearly shown with buildings clustered around the market places which remained their economic heart, most of which – to judge from their shapes – had originated in the Middle Ages as areas of common land. Norwich was the largest urban centre, although it was still largely contained within its medieval walls, so that even the Norfolk and Norwich hospital was surrounded by open land. Then, as now, the city stood like a spider at the centre of a web of roads which radiated outwards through the county. It is surprising that Faden failed to provide a large-scale plan of the city, as he did

– on the margins of the map – for the towns of Yarmouth, King's Lynn, and (more surprisingly) Swaffham. Lynn and Yarmouth were major international ports: Swaffham may have been included because it was an important centre for 'polite society' in the west of the county. The Crown Inn and the Assembly Rooms are shown prominently on the plan provided, along with the parish church, market cross and parsonage.

Although vast amounts of goods and people were moved by road in late eighteenth-century Norfolk, it was still easier to transport bulky cargoes by water. Norfolk, largely bypassed by the industrial revolution, did not experience the kind of large-scale frenzy of canal construction seen in the Midlands and the North of England in the course of the eighteenth century, but several of its inland waterways were navigable, a number of significant improvements were made to them, and large amounts of traffic also passed along the county's coasts, moving goods to and from London and the north-east of England. In the west of the county the principal Fenland watercourses – the Great Ouse, the Old and New Bedford Rivers, the Well Creek, Popham's Eau, the Old Nene and the Middle Level Drain – were all regularly used by small boats, principally wherries and keels, as were a number of the main rivers feeding into these, running off the 'uplands' to the east. The Little Ouse was navigable as far inland as Thetford, following the construction of flash-locks or 'staunches', and other improvements, instituted by an act of 1669–70. The Wissey was navigable as far as Oxborough and the Nar – following improvements in 1759 – as far as West Acre.[7] Faden's treatment of these rivers is not very detailed. No locks are shown on the Nar or Ouse, and no attempt is made to indicate which of the numerous Fen watercourses were navigable. A few landing places are shown, however – the 'hithes' at Middleton, Methwold, Downham Market and Oxborough – but others known to have existed at the time, such as Otringhithe on the Little Ouse, are not marked.

On the other side of the county the principal Broadland rivers were all navigable, as they are today: the Yare as far as Norwich, the Chet to Loddon, the Thurne to a point above Potter Heigham and the Ant as far as Dilham. The Waveney was improved between Beccles and Bungay in the 1660s, with the construction of three locks, two of which ('Boterys' and Shipmeadow) are shown by Faden.[8] The Bure was also improved, for 15km above Coltishall – as far as the town of Aylsham – following a parliamentary act of 1773. The work began in 1774 and the 'Aylsham Navigation' was completed by 1779.[9] Again, Faden shows the various locks that were constructed, at Aylsham, Burgh (two), Oxnead, Buxton, Horstead and Coltishall. The contrast in this respect with the map's treatment of the western rivers, where no locks appear, may be explained by the fact that these Broadland examples were true locks, more sophisticated and engineered than the simple 'staunches' used on the Nar and Ouse. They were particularly necessary in the case of the Bure because the river gradient was great enough to drive watermills, a number of which had been established before the 'improvements' were carried out. Faden shows that all the locks were located next to these, on by-pass channels: the same pools supplied water

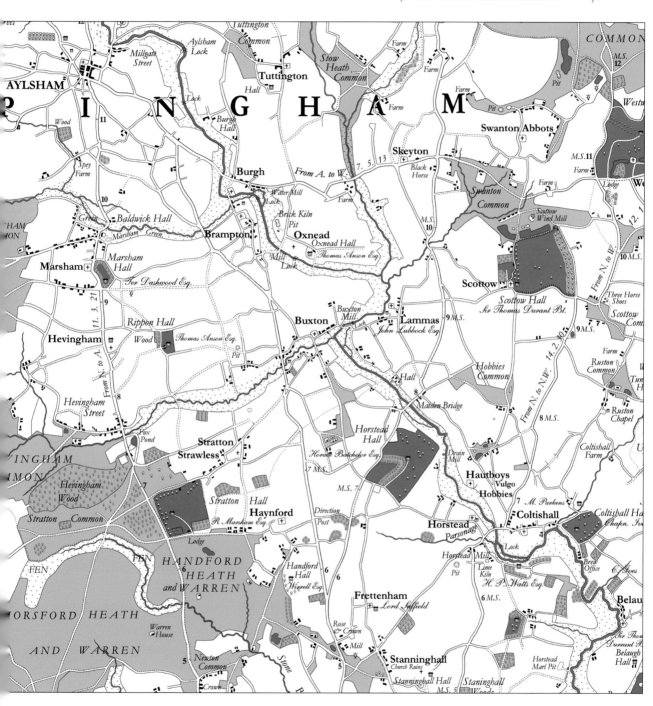

FIGURE 63. The Aylsham Navigation, showing locks and mills.

for both the waterwheels, and for refilling the locks (Figure 63).[10] Faden does not show the North Walsham and Dilham Canal, which extended navigation up the Ant by 14km, because this was only constructed in the 1820s. He does, however, show some of the numerous minor landing-places which served the

Broadland villages, here using the term 'staith' rather than 'hithe'. A number of 'staith houses', meaning minor warehouses where goods were stored prior to transportation, are also shown. A total of thirteen 'staiths' and 'staith houses' are shown in Broadland, many at the end of long, artificial cuts leading off the main rivers, some of which (like those leading to Tunstall and Geldeston Staithes) are labelled as 'Navigations'. This is only a fraction of the inland landing places in the district known to have been in existence at this time.[11] The different terms used on the map for small inland ports in different parts of Norfolk almost certainly reflects traditional usage. Hithe was an Old English term (*hid*), whereas staith was Scandinavian: one of several indications that the east and of the county, and to some extent the north, were more affected by Viking influence, and perhaps settlement, in the ninth century than the south and west.[12]

Major rivers, both in the east and east of the county, were arteries of communication. But they also represented barriers to movement by land, and as well as the numerous bridges carrying major roads across them Faden shows a number of ferries, again mainly in the Broads (Horning, Buckenham, Reedham, Whittlingham, Surlingham, Strumpshaw and Southtown near Yarmouth) but also in the west of the county (at Wilton, Oxborough, Welney, Southery, and Wilton, as well as at King's Lynn).

Coastal trade was much more economically important than that associated with the inland waterways, although the two were closely integrated, with Lynn in the west and Yarmouth in the east representing major transhipment points, where cargoes would be moved from sea-going vessels into the smaller keels and wherries. As well as these two great ports, medieval Norfolk had been provided with a number of others, some of which remained in operation in the late eighteenth century. The harbours at Wells and Blakeney are not specifically labelled on the map but the staithes at Brancaster, Burnham Overy and Holkham are marked, although the latter was no longer in use, and had merely become the name of the associated settlement: longshore drift, the build up of salt marsh, followed by embanking and reclamation had cut it off from the sea. Faden shows numerous other features of the coast associated with navigation. Hunstanton lighthouse, Cromer Light, the light houses at Happisburgh, the Winterton Ness Lights, and the Caister light houses, are all shown, smoke rising from them. Their purpose was not so much to warn ships off dangerous shallows as to provide aids to navigation. Low-lying, featureless coasts can take a ship's crew by surprise, and appropriate reference points are required to make them easier to 'read' from the sea. Lights provided seafarers with 'clearing lines or transits, alignments of nearer and further points in view which, when passed, permit a ship to turn in towards port or head off in another direction after passing a hazard', which explains why several of the examples shown by Faden come in groups.[13] In a number of other ways Faden seems to have given careful consideration to the way in which the coast would have appeared from out at sea. He thus marks the point where the 'mud cliffs' which dominate the north-east coast begin and end, the latter point also the start of the dunes, described by Faden as the 'North End of the Marrum Hills'. He marks the heights of the cliffs at several points. And, out at sea, he shows some of the main sandbanks and shallows, and provides sounding depths in fathoms. These features do not indicate that the map was actually used for coastal navigation – it was at once too expensive, and insufficiently detailed compared with contemporary naval charts. But they do reflect how in this period, when England's trade – both domestic and international – depended on shipping, and when British ships were the 'walls of oak' against French invasion, such matters were of interest to most educated people in a way that they would not be today. Faden carefully shows the batteries placed on the beach near Yarmouth, to defend that strategically important port from possible attack.

Industry

Norfolk in the 1790s was a predominantly rural county, but Faden's map does record a number of industrial activities. Chalk pits, marl pits, and 'pits' of unspecified purpose appear in a number of places (Figure 63), although Faden's surveyors did not attempt to record the hundreds, probably thousands, of small marl pits which by this time existed in the middle of arable fields in many parts of the county – everywhere, in fact, where acid soils overlay chalk or other calcareous deposits. Marling had been carried out since medieval times as a way of reducing soil acidity, but its scale increased markedly in the course of the seventeenth and eighteenth centuries, as enclosure and privatisation of land made it easier to carry out, as more and more acid, sandy land was brought into cultivation, and as new crops like turnips, particularly sensitive to 'sour' land, came into common use.[14] The chalk, which underlies most of the county, is only exposed or comes close to the surface in the west. Moving eastwards it dips gradually and becomes buried ever more deeply beneath more recent deposits, so that it is only exposed in the sides of the deeper valleys, most notably that of the Yare around Witchingham, and that of the Bure around Coltishall and Horstead.[15] Not surprisingly, Faden notes 'Horstead Marl Pit' beside the river, a little to the north of Wroxham Hall.

Although only two 'chalk pits' are specifically noted on the map, at Ringstead and Shernbourne, many of the other 'pits' recorded were probably for extracting chalk, for a high proportion lie in the north-west of the county, and comparatively few in the south-east, where the chalk is most deeply buried. Some of them may have supplied raw material for lime burning, although others perhaps were dug to obtain clunch, the hard chalk widely used in vernacular buildings (and, in the medieval period, for some high-status buildings, including monastic houses). Chalk was still sporadically used as a building material in Faden's day, usually in combination with flint and brick (notable examples include the early nineteenth-century Grove Farm in Marham).[16]

Lime was increasingly employed in the course of the nineteenth century for agricultural purposes, as a more efficient alternative to marl, but in Faden's day most kilns probably supplied the building industry – lime was used for making both mortar and plaster. It was also used in whitewash for walls, and in the tanning industry.[17] The distribution of the eighteen kilns which Faden depicts – very probably only a small proportion of those which actually existed at the time – appears to have principally been a function of the availability of chalk, and only to a lesser extent of the presence of suitable fuel supplies for the kilns, or of the proximity to markets (although the map resolutely fails to show the extraction pits which must have accompanied the kilns). Most examples were thus located in the north and west of the county. There were none in the south-east, while in the centre of the county they were restricted to the valleys of the major rivers, which cut through the overlying deposits to the chalk below. The presence of Norwich was probably another factor in their concentration in this part of the county, however, and examples are shown at Easton and Lakenham close to the Yar, immediately to the south of the city: and, slightly further afield, at Swannington in the valley of the Wensum and at Coltishall on the Bure.

Brick kilns appear to have been more numerous, and more widely scattered through the county, although once again how completely they were recorded by Faden's surveyors is uncertain (Figure 64). Small brickfields were numerous in Norfolk at the time, some operated by large estates, others by petty industrialists or part-time farmers, and Faden's surveyors probably only recorded the largest or most visually prominent examples. In this case, distribution was much less restricted by the availability of raw materials, for clays suitable for brickmaking can be found across much of the county.[18] While the boulder clays in the south and centre of the county can be used to make serviceable bricks, so too can the 'Norwich brickearth', the more sandy glacial drift found in the north-east of the county; the Kimmeridge clay in the far west; and even the alluvial clays found on the northern Fens and the Broads (Faden's map shows a kiln in the middle of Gaywood Marsh). The location of kilns was partly determined by local demand but mainly, perhaps, by the availability of fuel, for a very high proportion of those shown on the map are beside woods and, in particular, heaths, where gorse was readily available and made good kiln firing. Particularly noteworthy is the cluster of kilns at the northern end of Stock Heath, which may have encouraged the maintenance and management of the numerous pollards which, as we have seen, grew there (above, p. 126).

The most ubiquitous signs of industry on the map are wind and watermills (Figure 65). As already noted, a number of the drainage mills on the Norfolk Broads are shown with a windmill symbol, and some of these may have been dual-purpose structures, used both for grinding and drainage (there is good evidence that St Benet's mill in Broadland, for example, was used in part for milling rape seed, while in the following century, certainly, Dilham Dyke Mill in Smallburgh and Swim Coots Mill at Catfield were also used for grinding corn).[19] Most of the windmills shown were, however, unquestionably corn

FIGURE 64. The distribution of pits, kilns, and brickworks shown on Faden's map.

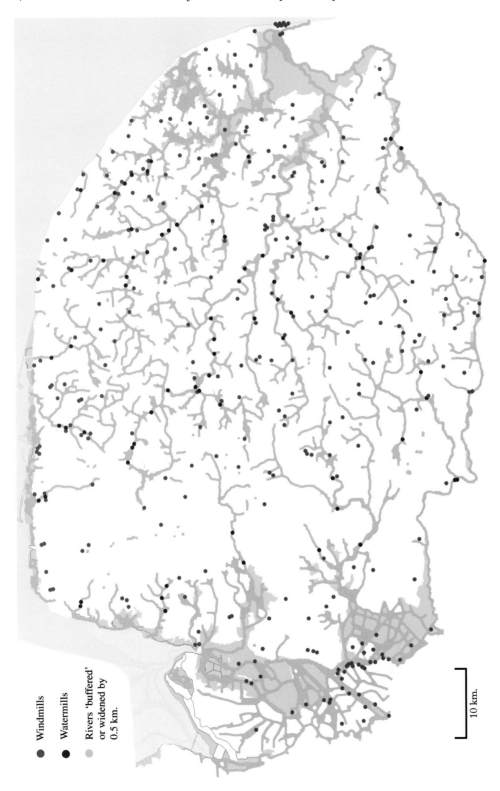

Windmills

Watermills

Rivers 'buffered'
or widened by
0.5 km.

10 km.

FIGURE 65. The location of wind and water mills, and their relationship to watercourses. Note how the majority of windmills avoid the higher interfluves, clustering instead close to the main valleys.

mills, for grinding grain to make flour or animal feed. With watermills the situation is more complex. Water, unlike wind, provided a continuous and relatively dependable source of energy which could be used to power a variety of industrial process, most notably paper-making, which in Norfolk dates back as an industrialised process to the seventeenth century. Some paper mills are specifically labelled on the map – on the Wensum at Bintry, Lyng, Swanton Morley, and Taverham – but it is possible that some of the other mills shown by Faden also served this purpose. It is notable, however, that even in the nineteenth century the majority of such enterprises were located on the middle stretches of this river, above Norwich.[20]

So far as the evidence goes, Faden's map provides a fairly comprehensive record of mills: both wind and watermills were prominent features of the landscape, hard to miss, which purchasers would have expected to see placed in their correct positions. No less than 274 windmills are shown (excluding examples which may have been used principally for drainage) and 70 watermills (excluding those specifically labelled as 'paper mill'). Watermills had gone through something of a design revolution in the course of the seventeenth and eighteenth centuries.[21] Earlier mills had generally been of undershot type, powered by the simple force of flowing water hitting the bottom of a water wheel. But during the seventeenth and eighteenth centuries wheels became larger and breast shot wheels, which turned as water filled long 'buckets', often replaced simple undershots: changes which could necessitate, in Norfolk's gentle topography, complex schemes of engineering in order to form larger supply ponds. Mills generally became more substantial and more sophisticated, often for example driving more than one set of stones. The distribution of watermills shown on Faden's map is interesting in a number of respects. They were not found along all stretches of the county's rivers. They were generally absent from the higher reaches, where there was often insufficient water, especially during the dry Norfolk summers. They were also generally absent from the lower reaches, mainly because here the rivers ran sluggishly, with little fall, but also because – in both the east and the west of the county – these stretches were often navigable. Where this was the case mills had to be by-passed, using locks – something which was carried out on some scale when the Aylsham Navigation was constructed in the 1770s, as we have seen, but which was expensive and generally avoided. For both these reasons, no watermills were to be found on the Bure below Horstead, none on the Yare below Keswick, and none below New Mills in Norwich on the Wensum; while in the west of the county, Thetford was the lowest mill of the Little Ouse, that at Northwold the lowest on the Wissey, and Pentney the lowest on the Nar. A few mills could be found in the lower reaches of rivers with outfalls on the north and west coasts, as at Burnham Overy, but this was exceptional. In this part of the county the slightly more eventful topography ensured that the gradients of watercourses were generally greater than elsewhere.

Windmills had also undergone, or were undergoing, a technological revolution at the time that Faden's map was surveyed. In the sixteenth and seventeenth centuries most if not all of those in Norfolk appear to have been post mills, in which the body of the mill, including the stones, could be turned on a large post in order to face the wind.[22] Such mills continued to be used, and indeed built, in the county during the eighteenth century, but expanding grain production meant that larger mills, capable of driving two or more sets of stones as well as machinery for cleaning grain and dressing the flour, were increasingly constructed. Tower mills of brick and smock mills of timber framing – in which only the cap was turned to face the wind – were thus replacing the older form at the time the map was surveyed, for these could be made taller and therefore generate more power.[23] During the late eighteenth century the fantail came into use, which turned the cap of the mill automatically into the wind, and this, too had probably been installed on some Norfolk mills by the time that Faden made his map.

The distribution of windmills is particularly interesting. Somewhat surprisingly, they do not usually stand on the highest ground in any district. Few, indeed, are to be found away from major valleys, on the wide interfluves between drainage systems. The reasons for this are not entirely clear. In part it may reflect the fact that on higher ground mills were more exposed to the elements: mill owners were acutely aware of the damage which could be caused to sails, cap or even the body of the mill by high winds. To some extent it may have been because the main centres of population were generally located in the valleys, rather than on the interfluves between them – although in Norfolk the extreme dispersion of settlement, especially in the south and east of the county, rather lessens the force of this argument. Certainly, in part it reflects another interesting aspect of mill location: the fact that watermills and windmills often stood close together. Indeed, in some places they were jointly labelled on the map, as in the case of 'Worstead Wind and Water Mills'. This proximity is probably because it made sense for millers to own one mill of each type in order to ensure that flour could still be ground even if there was no wind, or insufficient head of water to turn a wheel. It may be that the practice of owning two mills on proximate sites was becoming more common at this time, and at Wighton in the north of the county Faden notes 'Old Wind Mill' and, a few kilometres to the north-east, 'Water Mill and New Wind Mill'. Indeed, on rare occasions water and wind mill might be combined in the same building, as at Little Cressingham, which still survives in the south-west of the county, built in 1821. This replaced separate water and wind mills which were under the same ownership and advertised for sale together in 1795:[24] Faden shows them both, then separated by some 600m.

A significant number of windmills were located towards the middle of commons, a traditional location for industrial structures, convenient for the farms and cottages scattered along the common edge, but far enough away from them to reduce the risk of fire – mills were always burning down. In addition, windmills often clustered near the coast. Indeed, those at Cley or Burnham

Overy are still a familiar (and much-photographed) part of the Norfolk scene. Mills were prone to damage from sea gales, but coastal breezes were more reliable than inshore winds. The most notable concentration was on the beach at Yarmouth, where no less than seven stood in a line, although this also reflects, of course, the local market for flour provided by the town. At Wells on the north coast, similarly, four stood in close proximity.

Other signs of industry appear to have been unevenly, and rarely, noted by Faden and his surveyors. A number of 'Tan Offices' are marked – places where leather was tanned in pits filled with water and bark. Their concentration across the north of the county (with examples shown between Holt and Letheringsett, at Thurgarton, Guist, and just to the north of Blickling Park) may reflect the interests and biases of a particular surveyor, although this was one of the districts in the county where, to judge from other sources, the industry was important.[25] A 'Brew Office' is shown at Letheringsett, and a few examples of 'smith's shops', but why these examples, out of the hundreds which must have existed across the county at the time, were singled out for particular attention is unclear.

Other features

Faden shows a wide range of other features, the character and distribution of which would repay further investigation: we merely touch briefly on some of them here. Their presence on the map reflects in part their physical prominence in the landscape, and in part their prominence in the concerns of the map's potential purchasers. The late eighteenth century mind was, as we have argued, obsessed with the concept of 'improvement'. But people at the time were also keenly interested in what was seen as a potential threat to the continued amelioration of living conditions – demographic expansion and the growth in the numbers of the poor and unemployed. Faden's map was surveyed when the poor and indigent were still cared for under the terms of the 'Old Poor Law', created by a series of government interventions and parliamentary acts between 1563 and 1601, and updated by various subsequent administrative measures. Individual parishes were responsible for maintaining their own poor and sick and most relief took the form of direct payments.[26] However, the increasing scale of rural unemployment and poverty, and the escalating costs of relief, ensured that some parishes or groups of parishes (often in towns) began to erect 'poor houses' or 'work houses' to provide accommodation and employment, a course of action encouraged by an act of 1722. By 1776, twenty-four such institutions had been created in Norfolk.[27] Over the following quarter century more extensive provision was made as unemployment continued to rise, and by 1803 some 130 workhouses had been built. Some were for single parishes; some for groups of parishes; while others served single hundreds, or groups of 'incorporated' hundreds, which were set up either through individual parliamentary acts or – after 1782 – by magistrates working under the provisions of the Gilbert Act.

Faden's map was thus made at a time when workhouse provision was expanding rapidly, but it is clear that his surveyors did not attempt to record all the various institutions for housing and employing the poor to be found in the Norfolk landscape. A total of seventeen Houses of Industry, Poor Houses and Work Houses are clearly labelled (with a further two over the county boundary in Suffolk, at Flixton and at Shipmeadow near Bungay). Most were towards the east of the county. Not surprisingly, it was the larger institutions which were most comprehensively recorded. Of the five workhouses which had been erected for hundreds or 'incorporated hundreds', all are clearly noted on the map: at Rollesby (for the hundreds of East and West Flegg), Smallburgh (for Happing), Heckingham (Loddon and Clavering), Wicklewood (Forehoe), and Gressenhall (Launditch and Mitford). These were generally substantial, imposing structures which, in the case of Rollesby, Wicklewood and Gressenhall, still survive, although much altered over the subsequent years. In contrast, only two of the Houses of Industry established under the Gilbert Act, out of four in existence by this time, were recorded (although that at Bawdeswell appears to be shown, unlabelled). More striking is the fact that only nine of the parish houses, less than a fifth of those which probably existed by the mid 1790s, are explicitly labelled. The larger institutions, catering for hundreds or groups of hundreds, were more obvious than local parish houses, which were often little different in size and appearance to private dwellings. But in addition it may have been that these large modern buildings – precursors to the more numerous institutions soon to be established under the terms of the 1834 New Poor Law – seemed, to surveyors and other contemporaries, to point the way towards how the 'problem' of the poor should be dealt with in the future. It is noticeable that many of these buildings are clearly placed on islands cut out of common land, as at Wood Norton: as already noted (above, p. 117) the law allowed partial enclosure of common land for the benefit of the poor and the erection of poor houses. Indeed, a number of parish work houses, although not specifically labelled, can be identified as buildings standing on tiny plots, surrounded on all sides by common land.

The map shows other prominent features which Faden doubtless thought would be of interest to the public, such as the Cold Bath House at Ditchingham and the race course at Briston. The locations of eight 'duck decoys', and of two 'decoy houses', are also noted, mostly in the Fens in the west of the county. The map unquestionably omitted a number of others which we know were still in use at this time, especially in the Broads (as at Mautby, Hemsby, and Winterton). Decoys were a method of trapping wildfowl which had been introduced from the Low Countries in the seventeenth century and consisted of a number of curving 'pipes' (tapering channels covered by netting supported on a framework of hoops) which led off from an area of open water, each of which terminated in a detachable bow-net. Along one side of each pipe screens of wood and reeds were erected, arranged in overlapping fashion. Wildfowl were lured into the net by using a combination of tame decoy ducks and a dog called a 'piper', which would run around the screens, jumping over the low boards or 'dog jumps' placed between them. The wild fowl were attracted towards what must have looked like an appearing and disappearing dog and encouraged by the behaviour of the tame decoy ducks they swam towards it. When they had proceeded some way the decoy man would appear, driving the birds in flight down the pipe, and thus into the bow net at the end.[28] Decoys appear to have held had a perennial interest for the local gentry, much of whose time was spent raising and shooting game, and it is not surprising that a number are shown on the map, as they were on other eighteenth-century count maps.

Notes

1   B. P. Hindle, *Roads, Tracks and their Interpretation*. Batsford, London, 1993, pp. 64–6.

2   W. Albert, *The Turnpike Road System in England 1663–1840*. Cambridge University Press, Cambridge, 1972; Langford, *Polite and Commercial People*, pp. 391–408.

3   A. Davison and R. Joby, 'Early roads and turnpikes', in T. Ashwin and A. Davison, *An Historical Atlas of Norfolk*. Phillimore, Chichester, 2005, pp. 154–5.

4   *Ibid.*

5   C. W. Haines, 'Norfolk milestones, Part 3', *Journal of the Norfolk Industrial Archaeology Society* 4, 4 (1989), pp. 137–42.

6   Quoted in C. W. Haines, 'Norfolk milestones, Part 2', *Journal of the Norfolk Industrial Archaeology Society* 4, 2 (1987), pp. 46–52.

7   A. Davison, 'Inland waterways', in T. Ashwin and A. Davison (eds), *An Historical Atlas of Norfolk*, Phillimore, Chichester, 2005, pp. 156–7; J. Boyes and R. Russel, *The Canals of Eastern England*. David and Charles, Newton Abbot, 1977.

8   D. Best, 'The Bungay Navigation', *Journal of the Norfolk Industrial Archaeology Society* 7, 3 (2003), pp. 21–41.

9   Boyes and Russell, *Canals of Eastern England*, p.123.

10 Boyes and Russel, *Canals*, pp. 123–4.

11 Williamson, *Norfolk Broads*, pp. 137–9.

12 Smith, *Place Name Elements*, vol. 1, pp. 158, 278.

13 A. J. Parker, 'Maritime landscapes', *Landscapes* 2 (2001), pp. 122–41; p. 35.

14 W. M. Mathew, 'Marling in British agriculture: a case of partial identity', *Agricultural History Review* 41 (1993), pp. 97–110; Wade Martins and Williamson, *Roots of Change*, pp. 55–60.

15 B. Funnell, 'Geological background', in T. Ashwin and A. Davison (eds), *An Historical Atlas of Norfolk*. Phillimore, Chichester, 2005, pp. 4–5; C. P. Chatwin, *British Regional Geology: East Anglia and Adjoining Areas*. HMSO, London, 1961.

16 N. Pevsner and B. Wilson, *The Buildings of England, Norfolk I: Norwich and North-East*. Yale University Press, London, 2002, p. 24.

17 J. Jones and M. Manning, 'Lime burning and extractive industries', in T. Ashwin and A. Davison (eds), *An Historical Atlas of Norfolk*. Phillimore, Chichester, 2005, pp. 170–1.

18 R. Lucas, 'The brick kilns of Norfolk' and 'Brickmaking', in T. Ashwin and A. Davison (eds), *An Historical Atlas of Norfolk*, Phillimore, Chichester, 2005, pp. 162–3.

19 H. Apling, *Norfolk Corn Windmills*. Norfolk Windmills Trust, Norwich, 1984, pp. 276–7; Williamson, *Norfolk Broads*, p. 133.

20 M. Fewster and M. Scott, 'Paper mills and watermills', in T. Ashwin and A. Davison (eds), *An Historical Atlas of Norfolk*. Phillimore, Chichester, 2005, pp. 166–7.

21   J. Vince, *Discovering Watermills*. Shire, Aylesbury, 1987; Fewster and Scott, 'Paper mills and watermills'.

22   Apling, *Norfolk Corn Windmills*; Hills, *Power from Wind*; A. Yardy and M. Scott, 'Windmills', in T. Ashwin and A. Davison (eds), *An Historical Atlas of Norfolk*. Phillimore, Chichester, 2005, pp. 172–3.

23   Hills, *Power from Wind*, pp. 50–79.

24   J. M. Shaw, *Little Cresingham Mill*. Norfolk Windmills Trust, Norwich, 1982, p. 5.

25   C. Barringer, 'Tanners and tanning', in T. Ashwin and A. Davison (eds), *An Historical Atlas of Norfolk*. Phillimore, Chichester, 2005, pp. 160–1.

26   J. Crowley and A. Reid (eds), *The Poor Law in Norfolk, 1700–1850*. EARO, Ely, 1983; P. Slack, *The English Poor Law, 1531–1782*. Macmillan, London, 1990; D. Dymond, 'Workhouses before 1834', in P. Wade Martins (ed.), *An Historical Atlas of Norfolk*. Norwich, 1993, pp. 142–3.

27   Dymond, 'Workhouses before 1834'.

28   R. Payne-Gallway, *The Book of Duck Decoys*. Van Voorst, London, 1886; R. E. Baker, 'Norfolk duck decoys', *Transactions of the Norfolk and Norwich Naturalists' Society* 27, 1 (1985), pp. 1–8; T. Southwell, 'Norfolk decoys', *Transactions of the Norfolk and Norwich Naturalists' Society* 2 (1879), pp. 538–55.

CHAPTER 8

Traces of Antiquity

Introduction

The discussion so far has largely concentrated on what we can learn from Faden's map about the landscape of late eighteenth-century Norfolk. But, as intimated in Chapter 4, the map can also be used in another way: to provide evidence about much earlier landscape history. This comes in two forms. The first, interesting enough but of lesser importance to modern historians, is the direct representation of antiquities which the map includes: the sites of castles, Roman remains and the rest. More significant is the way that the configuration of commons, woods, and major boundaries shown on the map can be 'read' as evidence of the evolution of settlement, land use and territorial organisation in the remote, Anglo-Saxon past. In this chapter we will deal briefly with both these aspects.

Antiquities

Educated people in the later eighteenth century were keenly interested in the traces of antiquity in the landscape, although with archaeology in its infancy these were often misunderstood and misdated. In particular, Faden's map displays the common contemporary tendency – in part born of the importance of the classics in eighteenth-century culture – to attribute most substantial archaeological features, especially earthworks, to the Romans. Thus while the three major Roman sites in the county are correctly labelled – the Saxon Shore forts of *Branodunum* (Brancaster) and *Garrionenum* (Burgh Castle), and the provincial capital of *Venta Icenorum* at Caistor, just to the south of Norwich – a further six places are wrongly described as 'Roman Stations': Narborough, Attleborough, New Buckenham, Castle Acre, Castle Rising, and Thetford, the last further described as *Sitomagus*, a small settlement mentioned in the fourth-century document known as the *Antonine Itinerary*, which was well known at the time. In the case of the last four of these places the erroneous identification arose from a misidentification of substantial medieval earthworks associated with Norman castles: as Faden also shows and labels the castles themselves he presumably believed (together with a number of antiquarians, both before and after him) that Roman earthwork defences were frequently reused in the Middle Ages as castle sites.[1] In the case of Narborough, in contrast, it was the large Iron Age enclosure or 'hillfort' which was misidentified as a Roman monument. The identification of Attleborough, which features neither a medieval castle nor an Iron Age hillfort, as a Roman site may be due to the presence of the enigmatic linear earthwork called Bunn's Bank which begins just to the south of the town (although the only stretch of this feature which is actually shown and named on the map runs much further to the south-east). In addition to these places, Warbury Hill on the north coast, near Stiffkey, is described as a 'Roman Camp'. This is a more intriguing case. There are no obvious defences on the hill today but excavations carried out in the 1930s on a Bronze Age round barrow recovered Iron Age and Roman material, including masonry, while a stretch of wall, possibly of Roman date, was located here in 1985, leading to the suggestion that this may have been the site of a Roman watchtower, possibly part of the Saxon Shore defences.[2] Perhaps in Faden's day rather more evidence of Roman occupation survived above ground: the hill has been extensively quarried since.

Five Roman roads are also shown on the map. Two are unquestionably Roman – Peddars Way in the west of the county, and the Pye Road, the modern A140, in the south.[3] The others are more dubious, however. Thus the name 'Stone Street' is given to the B1332 Norwich-Bungay Road at Kirstead but, while the section of that road further to the south – between Woodton and Dichingham – may be Roman in date, the rest does not appear to be. The B1527 between Woodton and Hempnall is similarly described as a Roman road, but is not normally accepted as such, although it may well be an ancient route; while the name 'Icknield Way' given to the B113 Norwich to New Buckenham Road reflects an antiquarian tendency, widely shared in the eighteenth and nineteenth centuries, to continue the line of that supposed prehistoric trackway, running from Wessex, and along the escarpment of the Chilterns, as far as Norwich.[4]

Medieval remains are dealt with more competently. The sites of the main masonry castles are shown, with the exception of Mileham and Baconsthorpe, the latter presumably because the adjoining hall was still in occupation, and marked therefore as a residence. The principal medieval abbeys and priories – Beeston, St Benet's, Bromholme, Castle Acre, Coxford, Creake, Hickling, Langley, Pentney, Shouldham and Walsingham – are all noted, together with the site of St Olaves across the boundary in Suffolk. Parish churches are recorded with particular care, as we might expect: a total of 711 are shown, of which forty-eight are described as 'ruined' and four as 'demolished'. In a further two cases the 'scites' of churches which had presumably disappeared completely – at Thorpe Market and Oby – are noted. The numerous ruined and demolished churches are partly associated with deserted or shrunken settlements (such as Bayfield or Godwick), and partly result from the removal, in the period after the Reformation, of examples in villages (such as Shottesham or Becchamwell) which had possessed more than one church: ecclesiastical over-provision was a peculiarity of the county, resulting from its wealth, and idiosyncracies of social structure, in the early Middle Ages. Indeed, the 711 parish churches depicted – one for every 7.4 square kilometres – represents a density greater than that of any other county in England, although

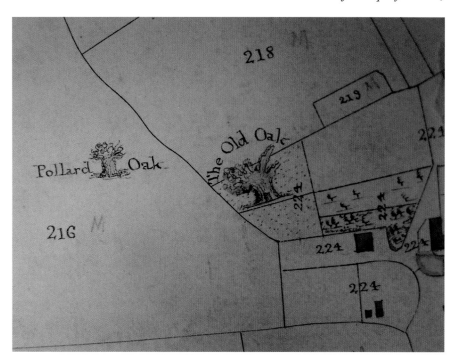

FIGURE 66. The 'Old Oak' at Winfarthing, as depicted on the tithe award map (courtesy Norfolk Record Office; photo by Rob Liddiard).

even this represents a decline from the situation in the high Middle Ages, when – if we include the numerous parishes in Norwich – there may have been as many as 928 in the county, or one for every 5.6 square kilometres.[5]

The county's few prominent prehistoric monuments receive surprisingly little attention. Two of its Iron Age forts – at Creake ('Burrow Dykes Ancient Camp') and Warham ('Barrow Hills Ancient Camp') – are noted, but those at Tasburgh and Holkham are passed over in silence while that at Narborough, as already noted, is identified as a Roman 'Station'. No examples of round barrows are recorded, and the various linear earthworks in the county are ignored.

All these features were included on the map because Faden knew that his educated customers would expect to find them there: points of interest in the landscape, and places to visit. Even Norfolk's most ancient tree, the Winfarthing Oak, is marked clearly as 'The Old Oak' (Figure 66). It was evidently already on the tourist route. A few decades later it was described both by John Claudius Loudon in his *Arboretum Britannicum* of 1842 and by Samuel Taylor in *Arboretum et Fruticetem Britannicum* of 1836: the latter described it as 'a mere shell, a mighty ruin, blasted to a snowy white, but it is magnificent in its decay'.[6] The tree finally died at the start of the twentieth century, although its remains were only finally cleared away in the 1950s.

Commons and territories

We have already noted that one of the most important and valuable aspects of Faden's map is the way that it shows with some accuracy the configuration of the extensive areas of common land which still remained in the county at the end of the eighteenth century, the layout of which had largely been established by the high Middle Ages. By examining commons in relation to other features of the landscape, human and natural, it is possible to learn something about the development of settlement before this date, in the middle and later Saxon period.

Of course, as we have also emphasised (above, pp. 101–2), commons had not remained completely unaltered since the early medieval period. Some had disappeared completely: others had been encroached on from their margins, or had had 'islands' taken out of them, often to serve particular purposes, such as the provision of sites for poor houses. But in addition to these obvious and relatively small encroachments, and shading into them, are much larger 'islands' – substantial tracts of farmed land ringed by continuous or near-continuous bands of common. Although these, too, can appear to be cut out of a once continuous tract of waste, this impression may often be an illusion, in the sense that such areas may simply represent tracts of favourable land, comparatively easy to cultivate, which are surrounded by areas of infertile or poorly-draining land better left for common grazing. The 'island', in other words, may always have been farmed, at least since the early Saxon period. But some examples do look as if they result from the fragmentation of more extensive tracts of grazing, which in most cases must have taken place before the eleventh or twelfth centuries, when the migration of settlement to the edges of commons tended to fix them in their present positions.

To understand the origins and early history of commons it is useful to consider their location in relation to the boundaries of the units called 'hundreds'. These were administrative subdivisions of shires or counties which were established in later Saxon times.[7] In the case of East Anglia, they were probably created when the region was reconquered from the Danes in the early tenth century and absorbed into the newly unified kingdom of England, under West Saxon rulers. Their boundaries were changed on a number of occasions in the course of the Middle Ages, and the discussion that follows relates not to the hundreds as shown (not very accurately) on Faden's map, but rather as we first encounter them, in the pages of Domesday Book.[8] Faden does not show us these, of course: they have to be reconstructed from the information in Domesday, and then 'imported' and laid onto the map.

Although hundreds as such were created relatively late in the Saxon period, their names sometimes suggest earlier origins, as ancient estates or folk territories. Some for example were named after what are probably early Anglo-Saxon tribes, such as Loddon and Clavering.[9] Most, however, seem to have taken their names from the places where the hundred court or 'moot' was held. These included ancient barrows (Greenhoe, Forehoe, Grimshoe, all incorporating the word *haugr*, Scandinavain for 'burial mound'); woods (Wayland Wood, which incorporates the Scandinavian term *lundr*, perhaps in the sense of 'sacred grove' – the traditional association of this place with the

'Babes in the Wood' story implies, perhaps, some ancient pagan significance);[10] and vills which functioned as major royal manors, such as Taverham or Holt. Several take their names from fords, which made good meeting places in cases where a major river ran through the centre of the hundred. Mitford is thus the 'moot ford', Depwade the 'deep ford'.

Not only the boundaries of hundreds, but also their meeting places, appear to have changed over time, and many are unknown or uncertain. But it is just possible that Faden provides clues to where some of the earliest may have been located. At nine locations he shows the location of gibbets, places where the bodies of executed criminals were hung in chains – usually on commons – as a warning to others.[11] Documentary and archaeological evidence suggests that many of these sites were very old, and that some had been used in the Middle Ages as actual places of execution – that is, they had originally been gallows sites. Some of these may in turn have developed from *cwealamstow*, places of pre-Conquest execution.[12] The latter were often located at or near hundred meeting places, allowing for the convenient dispatch of miscreants convicted at the hundred courts,[13] and in this context it is noteworthy that Faden's gibbets are not generally found near the margins of hundreds (although they often lay on the boundaries of individual parishes) but rather towards the centre of these territories, and thus perhaps indicating former moot sites (Figure 67). That in Holt hundred, for example, stood on common land in the east of Holt parish, roughly in the middle of the hundred; while that on Badley Moor, to the south-east of Dereham, is again more or less central to the hundred of Mitford, close to the infant river Tiffey, and it is tempting to suggest that the nearby ford was the meeting site from which the Hundred took its name. The gibbet on Methwold Heath, near to the parish boundary between Methwold and Feltwell, likewise lay close to the centre of Grimshoe Hundred, and a number of barrow sites are known from the immediate area. There are exceptions to this general rule, where gibbets are located asymmetrically within hundreds. Examples include the gibbet at West Bradenham, in the eastern part of Greenhoe Hundred; and that on the boundaries of the parishes of Kettletone and Ryburgh, in the western part of Gallow. But it is noteworthy that in both cases they stood in the centre of marked projections or promontories of the hundreds in question, and it is just possible that these represent older territorial units out of which hundreds had themselves been constructed in late Saxon times.

One sign that hundreds were ancient units, and not simply the arbitrary creations of medieval administrators, is the way that they often displayed a clear relationship with the natural topography. Their boundaries either followed major rivers like the Wensum, Thet or Bure, or else the watersheds lying between major drainage systems. They also displayed some association with the larger tracts of common land. Particularly striking is the line of the boundary between the hundreds of Holt and North Greenhoe, continued as that between Eynesford and Gallow, which followed a chain of commons for several miles on the watershed between the Glaven and the Stiffkey, and the

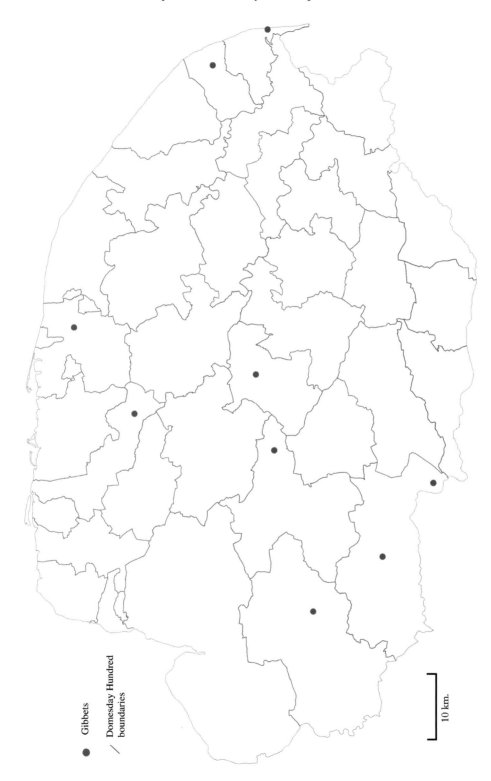

Gibbets

/ Domesday Hundred
boundaries

10 km.

FIGURE 67. Sites of gibbets shown by Faden, and their relationship with hundred boundaries.

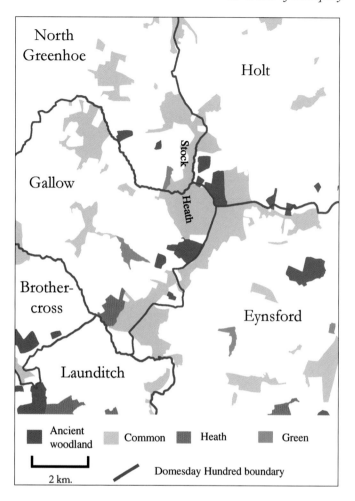

North
Greenhoe

Holt

Stock

Heath

Gallow

Brother-
cross

Eynsford

Launditch

Ancient
woodland | Common | Heath | Green

2 km.

Domesday Hundred boundary

FIGURE 68. Hundred
boundaries converging
on the great Stock Heath
in north Norfolk.

Glaven and the Wensum, passing through the middle of the vast Stock Heath. The boundary between Taverham and Blofield Hundreds similarly ran through the centre of Mousehold Heath. Extensive commons and heaths were, moreover, a particularly noticeable feature of the points where three hundreds met. The bounds of Shropham, Wayland and Grimshoe thus converged on the great area of heath around Wretham Warren, while Eynesford, Gallow, Holt and North Greenhoe all met on Stock Heath (Figure 68).

In this context, it is interesting to examine the long, continuous boundary which divided the hundreds of Forehoe and Humbleyard, Forehoe and Depwade, and Forehoe and Shropham, and which also marked, for much of its course, the boundary of the vast parish of Wymondham. This long line again followed an important watershed: that between the Rivers Tiffey and the Tas to the east, and between the Tiffey and the various tributaries of the Thet to the south and the west. The parishes based in the Tas valley form a series of parallel strips running westwards up to this high watershed boundary, onto the heavy soils of the Beccles Association, and the relationship of their boundaries to the

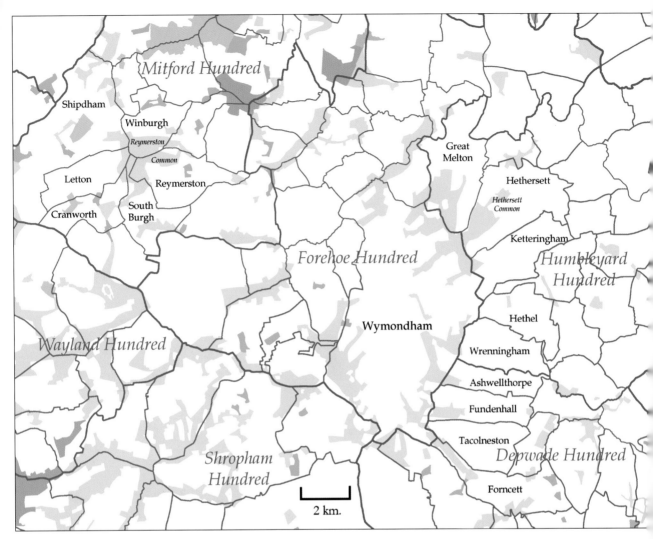

Within the figure:

Mitford Hundred

Shipdham

Winburgh

Reymerston
Common

Letton

Reymerston

South
Burgh

Cranworth

Great
Melton

Hethersett

Hethersett
Common

Ketteringham

Humbleyard
Hundred

Forehoe Hundred

Wymondham

Hethel

Wrenningham

Ashwellthorpe

Fundenhall

Wayland Hundred

Tacolneston

Depwade Hundred

Shropham
Hundred

Forncett

2 km.

common land shown on Faden's map makes it clear that the latter had once been much more continuous and extensive (Figure 69). To the south-east of Wymondham, the parish of Forncett St Peter has a long funnel-like projection which extends westwards up to the hundred boundary. This projection was entirely, at the time the map was surveyed, occupied by an area of common land: but this did not extend across the parish boundary to the north, into Tacolneston, for that parish had lost all its common land by the time the map was surveyed – it was enclosed by parliamentary act in 1778. The next parish along, however – Fundenhall – had not yet been enclosed at the time Faden's map was surveyed, and here once again the western end of the parish was occupied by common land. But not all of it. The common was located against the southern boundary of this portion of the parish, so that an area of enclosed land lay between it and the next parish boundary to the north. Immediately beyond that boundary, however, in Ashwellthorpe, common land began again, and there can be little doubt that the two areas would once have been continuous, for the boundary between the two parishes, as usual, does not mark any radical change in soils or topography. A similar pattern can be seen a little

FIGURE 69. The area around Wymondham, showing commons, hundred boundaries and parish boundaries.

further to the north, where Hethersett Common lay on the southwestern side of the parish, on heavy Beccles soils and abutting on the watershed boundary. Here again the common stopped abruptly at the parish boundary with Great Melton, and a broad strip of enclosed land lay beyond, with Melton's commons lying in a band beyond this: once again, the two areas must once have formed a continuous block of 'waste'. In fact, all the commons in the parishes around the periphery of Wymondham, running up to the hundred boundary, had a torn, tattered appearance, with clear discontinuities at parish boundaries, as did those extending along the watershed boundary running to the west. Extensive though they still remained in the 1790s, these upland commons had once – in Saxon times – clearly been more extensive and continuous.

In part the close relationship between hundred boundaries and commons was a simple function of topography. Because the former ran along high watersheds, occupied by heavy or infertile soils, they were bound to coincide with areas which were never brought into cultivation. But in part the relationship may reflect the fact that hundreds originated as economic and tenurial, rather than simply as administrative, entities: or at least, were constructed out of a small number of very large territories or estates, whose principal settlements and arable land had been located in the major valleys, where the better soils were found, but which had possessed tracts of woodland and grazing on the poorer upland soils, towards their peripheries. Most landscape archaeologists believe that in the middle Saxon period, when population densities were relatively low and England was divided into a number of small kingdoms, most of the land had been organised into extensive units like this, often described as 'multiple estates', which may themselves have developed from more ancient, tribal territories.[14] Early ecclesiastical territories were similarly extensive – large *parochiae* attached to *minsters* – before fragmentation in the course of the later Saxon period, as population increased, as large estates broke up into smaller local lordships, and as local lords, or groups of free men, erected new churches to serve smaller areas.[15] The vast parish of Wymondham may represent a rare survival of such an archaic unit. Parish boundaries, in a densely settled county like Norfolk, probably became fixed by the later twelfth century and the vast size of this parish must relate to the fact that the abbey was originally the site of a minster, and the parish was its parochia.[16] Quite why Wymondham should have survived in this manner is uncertain, but it is noteworthy that ecclesiastical integrity was paralleled in secular organisation. Dykebeck, a hamlet on the western edge of the parish, is listed in Domesday as a separate territorial unit (it comprised in fact two distinct holdings), while thirty free men in Wymondham itself were in the patronage of William de Warrenne. But the rest of the vill comprised a single huge manor, with a total population (including attached socmen) of 327 recorded individuals, implying a real population of at least 1,500. The manor had been held by Archbishop Stigand before the Conquest but by 1086 was in the hands of the King. It is possible that this was a return to an earlier situation, and that the place had been in royal hands *before* being given to Stigand. Certainly, in the thirteenth century the inhabitants of the town claimed the particular rights of tenants of ancient demesne of the Crown. Royal estates were less likely than others to fragment into numerous smaller territories, through sale or inheritance.

Wymondham was a rare survival. Other large territories, secular and ecclesiastical, had broken up into smaller units by the twelfth century. In areas where soils were both fertile and easily cultivated, these units were in general smaller than in districts where they were comparatively infertile and intractable. The former areas were able to sustain higher populations and thus a higher density of lords, and churches. This relationship is clear when the size of the county's many parishes is compared with the character of the soils within them (Figure 70). The smallest parishes in Norfolk, those extending over less than 300 hectares, are mainly concentrated on the fertile loams of the Wick 2 Association, and on the light clays of the Burlingham Association. They are seldom found on the heavier clays, or the more acid sands, areas where – as the parochial system was coming into existence – population densities were presumably rather lower. Parishes, although ecclesiastical units, generally perpetuated the outlines of one or more secular territories because parish churches were built by local lords or groups of freemen. For this reason their layout often displayed some relationship with soils and topography – their boundaries reflected a need to embrace the full range of resources required to maintain a farming community. Parishes, like the larger and more ancient territories in part 'ghosted' by the hundreds, thus show an intimate relationship with the areas of common land shown on Faden's map. Not only were many commons shared between more than one parish, as we have seen, but parish boundaries are often arranged in such a way that they seem to distort, stretch and extent in order to obtain a share of a particular area of common, indicating that it was a valuable resource. The way that Letton, Cranworth, South Burgh, Reymerston, Whinburgh and Shipdham all funnel in towards Reymerston Common is a particularly good example, although there are many others (Figure 71). Sometimes such arrangements seem out of all proportion to the small extent of the common in question. The parishes of Helhoughton, West Raynham, Weasenham St Peter, East and West Rudham, Harpley and Little Massingham thus all converge, or nearly converge, on the diminutive Keeping Heath, but this is probably the remnant of a once more extensive tract of common grazing, for the hundreds of Gallow, Launditch and Freebridge also meet near here.

Woods, wood pastures and commons

We noted earlier that one of the most striking features of Faden's map is that it reveals that a significant minority of commons in Norfolk still carried some tree cover in the later eighteenth century. There can be little doubt that more would have done so a century earlier and indeed, seventeenth-century maps show a number of commons as wooded which Faden shows as completely open, such

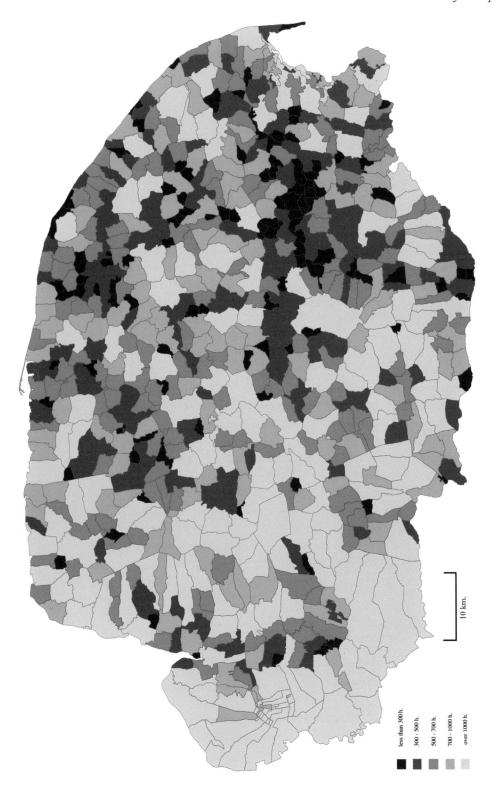

FIGURE 70. Variations in the size of medieval parishes in Norfolk. (Where two or more parishes have been amalgamated in post-medieval times, and their original boundaries unknown or uncertain, their area has been estimated). Based on Kain and Oliver (2001).

10 km.

less than 300 h.
300 - 500 h.
500 - 700 h.
700 - 1000 h.
over 1000 h.

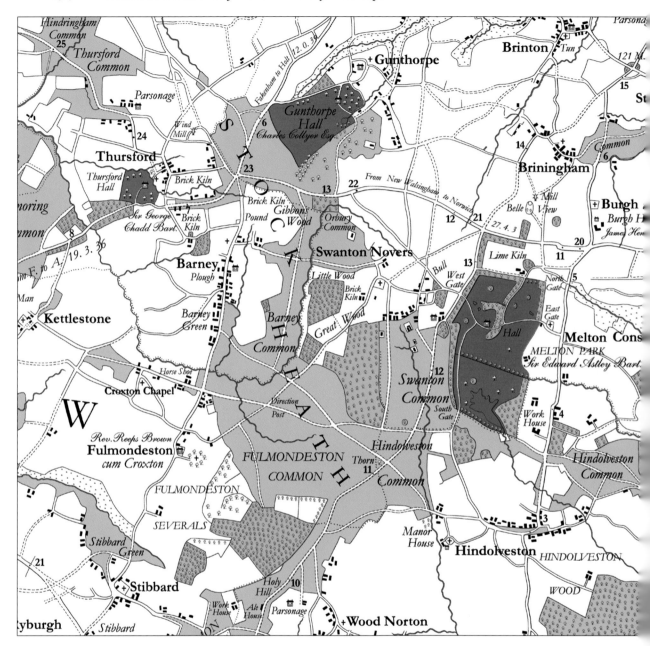

as the significantly named Hook Wood Common in Morley, which Waterman's map of 1629 depicts as almost entirely covered in trees.[17] It is probable that *most* commons, at least those on the heaviest or most acidic soils, had been wooded in the early Middle Ages, suggesting in turn that in many areas ancient woods, and commons, must have developed from the same extensive tracts of early woodland. While woods were enclosed and brought into complete private ownership by manorial lords in the eleventh, twelfth and thirteenth centuries,

FIGURE 71. Stock Heath, Melton Constable park, and neighbouring commons and wood pastures, as shown on Faden's map.

and were then managed more intensively as coppices, commons continued to be used as wood pastures, for grazing as well as for the production of wood and timber, and were thus gradually, in many cases, denuded of trees. Private wood pastures, of which Faden shows a number of examples, also presumably arose from the partial privatisation of these ancient wooded tracts in medieval times, although some may represent intakes from commons made at a relatively late date. In the Middle Ages the most important and most socially prestigious form of private wood pasture was the deer park. Parks were venison farms and hunting grounds which were usually located at some distance from the homes of their owners.[18] Only one park of this type still existed when Faden's map was surveyed, at North Elmham, although a few others had developed into the more ornamental parks which, from late medieval times, had become the accepted setting for large mansions (above, p. 129).

Careful scrutiny of Faden's map reveals that a high proportion (39 per cent) of ancient woods abutted directly upon a common. In many cases, moreover, woods, private wood pastures, wooded commons and open commons formed large, contiguous clusters. One good example can be found in the area between Hindolveston, Fulmondestone, Melton Constable, Barney, Swanton Novers, Gunthorpe and Thursford (Figure 71). The landscape here was dominated by the vast, rambling area of Stock Heath and its associated commons – Fulmondeston Common, Hindolveston Common, and Orbury Common. Attached to these were a number of areas of ancient woodland – including Swanton Novers Great Wood – together with an extensive block of private wood pasture the name of which – Fulmondeston Severals – suggests had been cut out of the adjacent common land at some point in the past. To the east, abutting directly on Swanton Common, was the great park surrounding Melton Constable Hall. Although, as we have said, the majority of parks shown by Faden originated in the course of the eighteenth century, and were created at the expense of farmland, Melton Park is one of the exceptions, originating as a medieval deer park. Like the various areas of ancient woodland, it had evidently been carved out of a single vast tract of wooded 'waste' in the early Middle Ages. In a similar way, Gunthorpe Park to the north appears likewise to have been carved out of Stock Heath at some unknown point in the past.

Similar, if less extensive, clusters of woods, commons and wood pastures can be found elsewhere. To the west of North Elmham, for example, Elmham Park, Elmham Heath, Brisley Green, and Horningtoft Wood must originally have formed one great block of woodland, separated by only a kilometre or so from another concentration, comprising Colkirk Common, Testerton Wood, and Ryburgh Wood, together with a further large tract of (unnamed) private wood pasture. Such arrangements are replicated, on a smaller scale, at many places. Typical is the close association of Hockering Wood, Swinnow Wood, Hockering Common, Stiphens Green (shown by Faden as still partly wooded), and the various unnamed areas of nearby woodland. The medieval park of Hockering, long vanished by the time the map was surveyed, had also once formed part of this cluster. It is noteworthy that in all these cases, and most others, these clusters of woods, wood pastures and commons occur on or close to hundred boundaries, and often at or near the point where three hundreds converge.

It is interesting to compare the distribution of private wood pastures, commons which still retained some tree cover, and ancient woodland shown on Faden's map, with that of the larger areas of woodland recorded by Domesday Book (Figures 72 and 73). As Patsy Dallas has noted, the two distributions are strikingly similar, both forming what Rackham has termed a 'woodland crescent' running in a great arc from the north-east of Norfolk, around to the claylands in the south of the county.[19] In part this pattern is structured by soils and geology, for much of this broad band corresponds with areas of poorly draining clays or – especially in the area to the north of Norwich – acid sands and gravels. But it is also a consequence of topographic factors, for this 'arc' also follows – more or less – the line of the great central watershed separating those Norfolk rivers which drain north and west into the North Sea from those – the Bure, Yare, Waveney, Wensum and their various tributaries – draining east, through the Norfolk Broads and Yarmouth.[20] Just as high watersheds often seem to have formed the boundaries between ancient territories based in river valleys, so this central watershed comprised a tract more generally marginal to the principal cultivated areas within the county in early times.

Although William Faden's map was surveyed in a relatively recent period, the fact that it predates the enclosure of most of Norfolk's commons means that it can tell us much about the development of the landscape at a much earlier period of time. In it we can see, albeit indistinctly, the existence of large Anglo-Saxon territories, fringed by extensive tracts of upland grazing; their fragmentation into smaller units during later Anglo-Saxon times; and the subsequent redistribution of settlement to the margins of the tattered remnants of the wooded pastures which had been left by the expansion of cultivation. Such things are not, however, revealed by examining the map alone, but only when we superimpose on it information from other data sets, especially those relating to administrative and ecclesiastical boundaries, and soils. Only by doing this, using modern digital technology, is it possible to glean the maximum amount of information about the history of the Norfolk landscape from Faden's fine map.

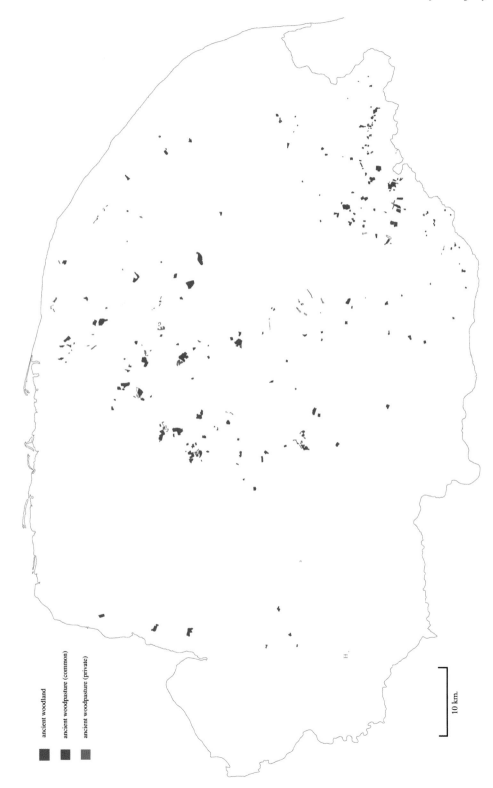

ancient woodland

ancient woodpasture (common)

ancient woodpasture (private)

10 km.

FIGURE 72. The distribution of all private wood pastures, wooded commons and probable ancient woods shown on Faden's map.

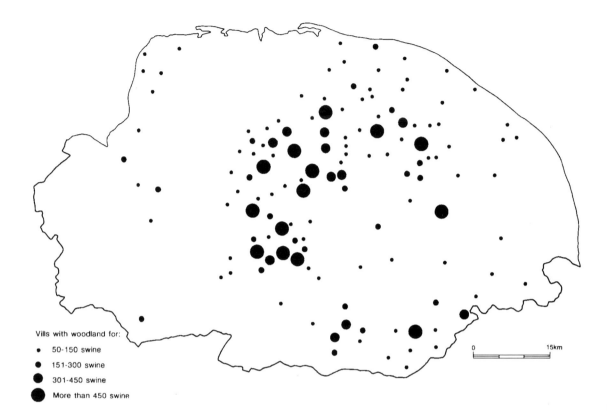

Vills with woodland for:

- · 50–150 swine
- ● 151–300 swine
- ● 301–450 swine
- ● More than 450 swine

0   15km

Notes

FIGURE 73. The distribution of woodland in 1066, as recorded in Domesday Book.

1   In the case of Thetford he was nearly correct: the castle here was built within Iron Age defences.

2   Norfolk Historic Environment Record No. 1865.

3   D. Gurney, 'Roman Norfolk', in T. Ashwin and A. Davison (eds), *An Historical Atlas of Norfolk*. Phillimore, Chichester, 2005, pp. 28–8; 'The Viatores', *Roman Roads in the South East Midlands, I*. Gollancz, London, 1964.

4   S. Harrison, 'The Icknield Way: some queries', *Archaeological Journal* 160 (2003), pp. 1–22.

5   N. Batcock, *The Ruined and Disused Churches of Norfolk*. East Anglian Archaeology 51, 1991, p. 1.

6   J. Grigor, *The Eastern Arboretum*. London, 1841, p. 354; T. F. Amyot, 'The Winfarthing Oak', *Transactions of the Norfolk and Norwich Naturalists' Society* 2, 1 (1875), pp. 12–18.

7   H. Cam, *The Hundred and the Hundred Rolls*. Methuen, London, 1930, pp. 10–11; H. Loyn, *The Governance of Anglo-Saxon England*. Edward Arnold, London, 1984, pp. 140–3; T. Williamson, *The Origins of Norfolk*. Manchester University Press, Manchester, 1993, pp. 126–33.

8   C. Barringer, 'Norfolk hundreds', in T. Ashwin and A. Davison (eds), *An Historical Atlas of Norfolk*. Phillimore, Chichester, 2005, pp. 96–7.

9   In Domesday the name of Loddon Hundred appears as *Lodningas*, 'the people of the *Ludne*', an early name for the River Chet; Clavering appears as Cnaveringas, 'the people

of *Cnava*'. Both seem to have originated as early Saxon tribal names. Williamson, *Origins of Norfolk*, pp. 64–5; 128–9.

10   Rackham, *History of the Countryside*, p. 79.

11   N. Whyte, *Perceptions of the Norfolk Landscape c.1500–1750*. Unpublished PhD thesis, University of East Anglia, 2005, p. 140.

12   A. Reynolds, 'The definition and ideology of Anglo-Saxon execution cemeteries', in D. Boe and F. Verhaeghe (eds), *Death and Burial in Europe*. Zelik, Bruges, 1997, pp. 33–41.

13   M. Carver, *Sutton Hoo: Burial Ground of Kings?* British Museum Press, London, 1998, pp. 137–54.

14   Williamson, *Origins of Norfolk*, pp. 83–92; P. Sawyer, 'Medieval English settlement: new interpretations', in P. Sawyer (ed.), *English Medieval Settlement*. Edward Arnold, London, 1979, pp. 1–8; G. Jones, 'Multiple estates and early settlement', in P. Sawyer (ed.), *English Medieval Settlement*, Edward Arnold, London, 1979, pp. 9–34.

15   J. Blair, *The Church in Anglo-Saxon Society*. Oxford University Press, Oxford, 2005.

16   T. Williamson, 'The landscape', in P. Cattermole (ed.), *Wymondham Abbey: A History of the Monastery and Parish Church*. Wymondham Abbey Book Committee, Wymondham, 2007, pp. 172–85; p. 173–4.

17   NRO, PD3/111.

18   R. Liddiard, *The Medieval Park: New Perspectives*. Windgather, Oxford, 2007.

19   Dallas, *Wood Pasture in Norfolk*, pp. 39–40; Rackham, 'Ancient woods of Norfolk', pp. 161–77; p. 165.

20   Williamson, *Origins of Norfolk*, pp. 14–17.

Conclusion

William Faden's map of Norfolk is unquestionably the most important single source of evidence we have for the landscape history of the county, not only in the eighteenth century but also in more ancient times. In part this is because it is one of the best of the maps produced by the eighteenth-century 'cartographic revolution'. But in part it is due to the fact that – unlike many of the county maps – it was surveyed before the great waves of parliamentary enclosure which removed many long-established details of the landscape, especially the configuration of commons and 'wastes', which had survived, largely unaltered, since the early Middle Ages. This said, many other county maps produced before the advent of the Ordnance Survey are potentially of equal value in understanding the history of particular areas. They can reveal not only the character of the physical landscape – the extent and distribution of such things as parks, woods, and industrial activities – but also, perhaps more significantly, the kinds of features which mapmakers and their customers considered to be important. Maps thus provide an insight into the mental worlds of the past, as much as the physical; and as with any historical source, we can only fully understand the information they present to us if we make the effort to understand the circumstances of their production. Above all, as we hope to have demonstrated, the amount and quality of information which the county maps can yield is vastly increased by redrawing them in digital form, and combining and comparing them with other digital datasets. We hope that our own interrogation and interpretation of William Faden's fine map of Norfolk will inspire others to undertake similar endeavours, in other parts of Britain.

Bibliography

Adams, G. (1791) *Geometrical and Graphical Essays Containing a General Description of the Mathematical Instruments used in Geometry, Civil and Military Surveying, Levelling and Perspective*. London.

Addington, S. (1982) 'Landscape and settlements in South Norfolk', *Norfolk Archaeology* 28, 97–139.

Ainslie, J. (1802) *The Gentleman and Farmers Pocket Companion and Assistant*. Edinburgh.

Ainslie, J. (1812) *A Comprehensive Treatise on Land Surveying Comprising the Theory and Practice in all its Branches*. Edinburgh.

Albert, W. (1972) *The Turnpike Road System in England 1663–1840*. Cambridge University Press, Cambridge.

Allison, K. J. (1957) 'The sheep-corn husbandry of Norfolk in the sixteenth and seventeenth centuries', *Agricultural History Review* 5, 12–30.

Amyot, T. F. (1875) 'The Winfarthing Oak', *Transactions of the Norfolk and Norwich Naturalists' Society* 2, 1, 12–18.

Apling, H. (1984) *Norfolk Corn Windmills*. Norfolk Windmills Trust, Norwich.

Bacon, K. (1993) *Enclosure in East Norfolk: Non-Parliamentary Enclosure and Consolidation in the Hundreds of Happing and Flegg*. Unpublished MA dissertation, School of History, University of East Anglia.

Bailey, M. (1988) 'The rabbit and the medieval East Anglian economy', *Agricultural History Review* 36, 1–20.

Bailey, M. (1989) *A Marginal Economy? East Anglian Breckland in the Later Middle Ages*. Cambridge University Press, Cambridge.

Bailey, M. (1990) 'Sand into gold: the evolution of the foldcourse system in west Suffolk, 1200–1600', *Agricultural History Review* 38, 40–57.

Baker, R. E. (1985) 'Norfolk duck decoys', Transactions of the Norfolk and Norwich Naturalists' Society 27:1, 1–8.

Balston, J. (1992) *The Elder James Whatman 1702–1759; England's Greatest Paper-Maker*. Privately published, West Farleigh, Kent.

Barnes, G. (2002) *Woodlands in Norfolk: A Landscape History*. Unpublished PhD thesis, School of History, University of East Anglia.

Barringer, J. C. (2002) *Bryant's Map of Norfolk in 1826*. Larks Press, Dereham.

Barringer, J. C. (1973) 'Introduction', in *Faden's Map of Norfolk*. Norfolk Record Society 42.

Barringer, J. C. (1989) 'Introduction', in *Faden's Map of Norfolk*. Larks Press, Dereham.

Barringer, J. C. (2005) 'Norfolk hundreds', in T. Ashwin and A. Davison (eds), *An Historical Atlas of Norfolk*, 2nd edn. Phillimore, Chichester, 96–7.

Barringer, J. C. (2005) 'Tanners and tanning', in T. Ashwin and A. Davison (eds), *An Historical Atlas of Norfolk*, 2nd edn. Phillimore, Chichester, 160–1.

Batcock, N. (1991) *The Ruined and Disused Churches of Norfolk*. East Anglian Archaeology 51.

Best, D. (2003) 'The Bungay Navigation', *Journal of the Norfolk Industrial Archaeology Society* 7:3, 21–41.

Birtles, S. (2003) *A Green Space Beyond Self-Interest: The Evolution of Common Land in Norfolk, c.750–2003*. Unpublished PhD thesis, University of East Anglia.

Blair, J. (2005) *The Church in Anglo-Saxon Society*. Oxford University Press, Oxford.

Blakemore, M. J. and Harley, J. B. (1980) 'Concepts in the History of Cartography'. *Cartographica* 17, (Monograph No. 26).

Blomefield, F. (1805) *An Essay Towards a Topographical History of the County of Norfolk*, 11 vols. London.

Boswell, J. (1823) *The Life of Samuel Johnson*. Glasgow.

Boulton, G. S., Cox, F., Hart, J. and Thornton, M. (1984) 'The glacial geology of Norfolk', *Bulletin of the Geological Society of Norfolk* 34, 103–22.

Boyes, J. and Russel, R. (1977) *The Canals of Eastern England*. David and Charles, Newton Abbot.

Bull, G. B. G. (1956) 'Thomas Milne's Land Utilisation Map of the London area in 1800', *Geographic Journal* 122, 25–30.

Bull, G. B. G. (1975–76) 'Introduction', in *Thomas Milne's Land Use Map of London and Environs in 1800*. London Topographical Society publications 118 and 119, London.

Burns, A. (1771) *Geodaesia Improved or a New and Correct Method of Surveying*. Tarporley, Chester.

Cam, H. (1930) *The Hundred and the Hundred Rolls*. Methuen, London.

Campbell, B. M. S. (1981) 'Commonfield origins – the regional dimension', in T. Rowley (ed.), *The Origins of Open-Field Agriculture*. Croome Helm, London, 112–29.

Campbell, B. M. S. (1981) 'The extent and layout of commonfields in east Norfolk', *Norfolk Archaeology* 28, 5–32.

Carroll, R. A. (1996) *The Printed Maps of Lincolnshire 1576–1900: Carto-Bibliography, with an Appendix on Road-Books 1675–1900*. Lincoln Record Society 84, Woodbridge.

Carver, M. (1998) *Sutton Hoo: Burial Ground of Kings?* British Museum Press, London.

Chambers, B. (1964) 'M. J. Armstrong in Norfolk: the progress of an eighteenth century county survey', *Geographical Journal* 130:3, 427–31

Chambers, B. (1983) 'Introduction', in *The County of Bedford, Surveyed Anno MDCCLXV, and Engraved by Thomas Jefferys, Geographer to His Majesty*. Bedfordshire Historical Record Society.

Chapman, J. (1987) 'The extent and nature of parliamentary enclosure', *Agricultural History Review* 35, 25–35

Chapman, J. and Seeliger, S. (2001) *Enclosure, Environment and Landscape in Southern England*. Tempus, Stroud.

Chatwin, C. P. (1961) *British Regional Geology: East Anglia and Adjoining Areas*. HMSO, London.

Chrimes, M. (2009) 'Society of Civil Engineers', in *The Oxford Dictionary of National Biography*, Oxford University Press, on line edition.

Chubb, T. (1928) *A Descriptive List of the Printed Maps of Norfolk 1574–1916*. Jarrolds, Norwich.

Clark, A. J., Tarling, D. H. and Noel, M. (1988) 'Developments in archaeomagnetic dating in Britain', *Journal of Archaeological Sciences* 15, 645–67

Close, C. (1969 originally published 1926) *The Early Years of the Ordnance Survey*. Repr. David and Charles, Newton Abbot.

Crerar, J. W. and Ross, J. A. (1953) 'John Dalton, Captain Joseph Huddart and the Harris family', *British Journal of Ophthalmology* 37:3, 181–4.

Crittall, E. (1952) 'Introduction', in *Andrews' and Dury's 1773 Map of Wiltshire*. Wiltshire Archaeological and Natural History Society 8.

Crompton, G. and Sheail, J. (1975) 'The historical ecology of Lakenheath Warren in Suffolk, England: a case study', *Biological Conservation* 8, 299–313.

Crone, G. R. (1951) 'Further notes on Braddock Mead, alias John Green'. *Imago Mundi* 8, 69–70.

Crone, G. R. (1978) *Maps and their Makers*. Dawson, Folkestone.

Crowley, J. and Reid, A. (eds) (1983) *The Poor Law in Norfolk, 1700–1850*. EARO, Ely

Dallas, P. (2005) *Wood Pasture in Norfolk: The Relationship between Trees, Animal Husbandry and the Changing Agricultural Landscape from the First Millennium AD to the Age of Improvement*. Unpublished MA dissertation, University of East Anglia.

Darby, H. C. (1983) *The Changing Fenland*. Cambridge University Press, Cambridge.

Darlington, I. and Howegego, J. (1964) *Printed Maps of London, circa 1553–1850*. George Philip, London.

David, A. (2003) 'Lt. Murdoch MacKenzie and his survey of the Bristol Channel and the south coast of England', *Cartographic Journal* 40, 1, 69–78.

Davison, A. (1990) *The Evolution of Settlement in Three Parishes in South East Norfolk*. East Anglian Archaeology 49.

Davison, A. (1994) 'The field archaeology of Bodney, and the Stanta Extension', *Norfolk Archaeology* 42, 57–79.

Davison, A. (2005) 'Inland waterways', in T. Ashwin and A. Davison (eds), *An Historical Atlas of Norfolk*, 2nd edn. Phillimore, Chichester, 156–7

Davison, A. and Cushion, B. (1999) 'The archaeology of the Hargham Estate', *Norfolk Archaeology* 53, 257–74.

Davison, A. and Joby, R. (2005) 'Early roads and turnpikes', in T. Ashwin and A. Davison, *An Historical Atlas of Norfolk*, 2nd edn. Phillimore, Chichester, 154–5.

Defoe, D. (1724) *A Tour Through the Whole Island of Great Britain*. London.

Delano-Smith, C. and Kain, R. J. P. (1999) *English Maps: A History*. The British Library, London.

Dymond, D. (1993) 'Workhouses before 1834', in P. Wade Martins (ed.), *An Historical Atlas of Norfolk*. Norwich, 142–3.

Dymond, D. P. (2003) 'Introduction', in *Joseph Hodskinson's Map of Suffolk in 1783*. Larks Press, Dereham.

Eden, P. (ed.) (1997) *Dictionary of Land Surveyors and Local Map Makers of Great Britain and Ireland 1530–1850*, 2 vols, 2nd edn by Sarah Bendall. British Library, London.

Ehrenberg, R. (2006) *Mapping the World: An Illustrated History of Cartography*. National Geographic Society, Washington.

Evans, I. and Lawrence, H. (1979) *Christopher Saxton: Elizabethan Map-Maker*. Wakefield Historical Publications, Holland Press.

Faden, W. (1963) *Catalogue of the Geographical Works, Maps, Plans &c.* Map Collectors Circle, London.

Fewster, M. and Scott, M. (2005) 'Paper mills and watermills', in T. Ashwin and A. Davison (eds), *An Historical Atlas of Norfolk*, 2nd edn. Phillimore, Chichester, 166–7.

Fisher, S. (2004) 'Joseph Huddart', in *The Oxford Dictionary of National Biography*. Oxford University Press, Oxford.

Fordham, H. G. (1929) *Some Notable Surveyors and Map Makers of the Sixteenth, Seventeenth and Eighteenth Centuries and their Work: A Study in the History of Cartography*. Cambridge University Press, Cambridge.

Fordham, H. G. (1925) *Daniel Paterson: His Maps and Itineraries 1771–1825*. British Library, 4th series, V, Vol. 4.

Fordham, H. G. (2003) *Studies in Carto-Bibliography*. Clarendon Press, Oxford, 1914; reprinted by Martino Publishing, Mansfield, CT, USA.

Fortescue, J. (1927) *The Correspondence of King George the Third from 1760 to December 1783*, 6 vols. Macmillan, London.

Foxell, S. (2007) *Mapping London: Making Sense of the City*. Blackdog Publishing, London.

Frostick, R. (2002) *The Printed Maps of Norwich 1558–1840: A Carto-bibliography*. Privately published, Norwich.

Frostick, R. (2007) 'James Corbridge – surveyor', *Norfolk Archaeology* 45:2, 155–75.

Funnell, B. (2005) 'Geological background', in T. Ashwin and A. Davison (eds), *An Historical Atlas of Norfolk*, 2nd edn. Phillimore, Chichester, 4–5.

Gough, R. (1780) *British Topography*, 2 vols. London.

Gregory, J. (2005) 'Mapping improvement: reshaping rural landscapes in the eighteenth century', *Landscapes* 6:1, 62–82.

Grigor, J. (1841) *The Eastern Arboretum*. London.

Grocott, T. (2002) *Shipwrecks of the Revolutionary and Napoleonic Eras*. Caxton Editions, London.

Gurney, G. (2005) 'Roman Norfolk', in T. Ashwin and A. Davison (eds), *An Historical Atlas of Norfolk*, 2nd edn. Phillimore, Chichester, 28–9.

Haines, C. W. (1987) 'Norfolk milestones, Part 2', *Journal of the Norfolk Industrial Archaeology Society* 4, 2, 46–52.

Haines, C. W. (1989) 'Norfolk milestones, Part 3', *Journal of the Norfolk Industrial Archaeology Society* 4, 4, 137–42.

Harley, J. B. (1963–64) 'The Society of Arts and the surveys of the English counties 1759–1809', *Journal of the Royal Society of Arts* 112, 43–6; 119–24; 269–75; 538–43.

Harley, J. B. (1964) 'William Yates and Peter Burdett: their role in the mapping of Lancashire and Cheshire during the late eighteenth century', *Transactions of the Historic Society of Lancashire and Cheshire* 115, 107–31.

Harley, J. B. (1965) 'The re-mapping of England 1750–1800', *Imago Mundi* 19, 56–67.

Harley, J. B. (1966) 'The bankruptcy of Thomas Jefferys: an episode in the economic history of eighteenth-century map-making', *Imago Mundi* 20, 27–48.

Harley, J. B. (1966) 'English county map-making in the early years of the Ordnance Survey: the map of Surrey by Joseph Lindley and William Crosley', *Geographic Journal* 132:3, 372–8.

Harley, J. B. (1966) 'From Saxton to Speed'. *Cheshire Round* 1, 174–84.

Harley, J. B. (1966) 'John Strachey of Somerset; an antiquarian cartographer of the early eighteenth century', *Cartographic Journal* 3:1, 2–7.

Harley, J. B. (1968) 'Error and revision in early Ordnance Survey maps', *Cartographic Journal* 5, 115–24.

Harley, J. B. (1968) 'Introduction', in William Yates, *A Map of the County of Lancashire 1786*. Historic Society of Lancashire and Cheshire, Birkenhead.

Harley, J. B. (1970) 'Cartographical introduction', in series *Reprints of the First Edition of the One-inch Ordnance Survey of England and Wales*, Norfolk sheets 37, 38, 45, 46 and 47. David and Charles, Newton Abbot.

Harley, J. B. (1976) 'George Washington: map maker', *Geographical Magazine* 48, 588–94.

Harley, J. B. (1979) 'Christopher Saxton and the first atlas of England and Wales 1579', *The Map Collector* 8, 3–11.

Harley, J. B. (1987) 'Introductory essay', in *The Old Series Ordnance Survey Maps of England and Wales, Vol. 5, Lincolnshire, Rutland, and East Anglia*. Harry Margary, Lympne Castle.

Harley, J. B. (2001) *The New Nature of Maps: Essays in the History of Cartography*. ed. Paul Laxton, John Hopkins University Press.

Harley, J. B. and Dunning, R. W. (1981) 'Introduction', in *Somerset Maps: Day and Masters 1782 and Greenwood 1822*. Somerset Record Society 76, Taunton.

Harley, J. B. and Harvey, J. C. (1973) 'Introductory notes', in *The county of York: Survey'd in MDCCLXVII, VIII, IX and MDCCLXX, Engraved by Thomas Jefferys, Geographer to His Majesty*. Harry Margary, Lympne Castle.

Harley, J. B. and Laxton, P. (1974) *Peter Burdett's Map of Cheshire, 1777: The Theory and Practice of an Eighteenth-Century County Survey*. Historic Society of Lancashire and Cheshire, Occasional series Vol. 1. Published with a facsimile and introduction by Lund Humphries.

Harley, J. B. and O'Donoghue, Y. (1975) 'Introduction', in *The Old Series Ordnance Survey Maps of England and Wales*, vol. 1. Harry Margary, Lympne Castle.

Harley, J. B. and Walters, G. (1978) 'English map collecting 1790–1840: a pilot survey of the evidence in Sotheby sale catalogues', *Imago Mundi* 2nd Series 30:4, 31–55.

Harley, J. B., Fowkes, D. V. and Harvey, J. C. (1975) 'Introduction', in *Burdett's Map of Derbyshire, 1791*. Derbyshire Archaeological Society, Derby.

Harley, J. B., Petchenik, B. B. and Towner, L. W. (1978) *Mapping the American Revolutionary War*. University of Chicago Press, Chicago.

Harris, L. E. (1953) *Vermuyden and the Fens: A Study of Cornelius Vermuyden and the Great Level*. Cleaver Hume Press, London.

Harrison, S. (2003) 'The Icknield Way: some queries', *Archaeological Journal* 160, 1–22.

Harting, J. E. (1898) *The Rabbit*. London.

Harvey, P. D. A and Thorpe, H. (1959) *The Printed Maps of Warwickshire 1576–1900*. Warwick County Council in association with the University of Birmingham.

Harvey, P. D. A. (1980) *The History of Topographical Maps: Symbols, Pictures and Surveys*. Thames and Hudson, London.

Hills, R. L. (1994) *Power from Wind: A History of Windmill Technology*. Cambridge University Press, Cambridge.

Hills, R. L. (2003) *The Drainage of the Fens*. Landmark Books, Ashbourne.

Hindle, B. P. (1993) *Roads, Tracks and Their Interpretation*. Batsford, London.

Hindle, P. (1998) *Maps for Historians*. Phillimore, London.

Hindle, P. (2001) 'The first large scale county maps of Cumberland and Westmorland in the 1770s', *Transactions of the Cumberland and Westmorland Antiquarian & Archaeological Society* 3rd Series 1, 139–53.

Hodge, C., Burton, R., Corbett, W., Evans, R. and Scale, R. (1984) *Soils and their Uses in Eastern England*. Soil Survey of England and Wales, Harpenden.

Hodson, Y. (1989) *Ordnance Surveyors' Drawings, 1789–c.1840*. Research Publications, Reading.

Huddart, J. (1796) 'Observations on horizontal refraction which effect the appearance of terrestrial objects, and dip and depression of the horizons of the sea', paper read to the Royal Society on 24 November 1796.

Huddart, W. (1989) *Unpathed Waters: The Life and Times of Capt. Joseph Huddart F. R. S. 1741–1816*. Quiller Press, London.

Hutton, C. (1778) 'An account of the calculations made from the survey and measures taken at Schiehallion, in order to ascertain the mean density of the Earth', *Philosophical Transactions of the Royal Society* 60, 689–788.

Jacques, D. (1983) *Georgian Gardens: The Reign of Nature*. Batsford, London.

James, N. (2000) 'The transformation of the Fens', in T. Kirby and S. Oosthuizen (eds), *An Atlas of Cambridgeshire and Huntingdonshire History*. Anglia Polytechnic University, Cambridge, 47–8.

Jones, G. (1979) 'Multiple estates and early settlement', in P. Sawyer (ed.), *English Medieval Settlement*. Edward Arnold, London, 9–34.

Jones, J. and Manning, M. (2005) 'Lime burning and extractive industries', in T. Ashwin and A. Davison (eds), *An Historical Atlas of Norfolk*, 2nd edn. Phillimore, Chichester, 170–1.

Kain, R. and Oliver, R. (2001) *Historic Parishes of England and Wales: An Electronic Map of Boundaries before 1850 with a Gazetteer and Metadata*, History Data Service, Colchester.

Kent, W. (1794) *General View of the Agriculture of Norfolk*. London.

Kingsley, D. (1980–81) *Printed maps of Sussex 1575–1900*. Sussex Record Society 72, Lewes.

Labarre, E. J. (1952) *Dictionary and Encyclopaedia of Paper and Paper-making*, 2nd edn. Amsterdam.

Lambert, J. M., Jennings, J. N., Smith, C. T, Green, C. and Hutchinson, J. N. (1952) *The Origins of the Broads*. Royal Geographical Society, London.

Langford, P. (1992) *A Polite and Commercial People: England 1727–1783*. Oxford University Press, Oxford.

Laxton, P. (1972) 'Introduction', in *County of Berkshire Map by John Rocque*. Harry Margary, Lympne Castle.

Laxton, P. (1976) 'Introduction', in *Two Hundred and Fifty Years of Map-making in the County of Hampshire: A Collection of Reproductions of Printed Maps Published Between the Years 1575 and 1826*. Harry Margary, Lympne Castle.

Laxton, P. (1976) 'The geodetic and topographical evaluation of English county maps 1740–1840'. *Cartographic Journal* 13, 37–54.

Liddiard, R. (2007) *The Medieval Park: New Perspectives*. Windgather, Oxford.

Livingston, D. N. and Withers, C. W. J. (eds) (1999) *Geography and Enlightenment*. Univeristy of Chicago Press, Chicago.

Loyn, H. (1984) *The Governance of Anglo-Saxon England*. Edward Arnold, London.

Lucas, R. (2005) 'The brick kilns of Norfolk' and 'Brickmaking', in T. Ashwin and A. Davison (eds), *An Historical Atlas of Norfolk*, 2nd edn. Phillimore, Chichester, 162–3.

Macnair, A. (2003) 'Foulsham and North Elmham: the landscape of two Norfolk parishes at the time of parliamentary enclosure', *Norfolk Archaeology* 44:2, 269–92.

Marshall, W. (1787) *The Rural Economy of Norfolk*. London.

Mathew, W. M. (1993) 'Marling in British agriculture: a case of partial identity', *Agricultural History Review* 41, 97–110.

Mingay, G. (1997) *Parliamentary Enclosure in England: Its Causes, Incidence and Impact*. Longman, London.

Monmonier, M. (1991) *How to Lie with Maps*. University of Chicago Press, Chicago.

Morley, K. E. (2003) *The Origins and Development of Common Land in the Boulder Clay Region of Norfolk*. Unpublished MA dissertation, School of History, University of East Anglia.

Mosby, J. E. G. (1938) *The Land of Britain: Vol. 70, Norfolk*. Geographical Publications, London.

Mudge, W. (1799, 1801 and 1813) *An Account of the Operation for Accomplishing a Trigonometrical Survey of England and Wales, 1797–1799*, 3 vols. London.

Mudge, W. and Dalby, I. (1799) 'An account of the operations carried on for accomplishing a trigonometrical survey of England and Wales 1784–1796', *Philosophical Transactions* 1, p. 411.

Mulkream, L. (ed.) (1949) *Thomas Pownall's Topographical Description of North America.* University of Pittsburgh Press, Pittsburgh.

Murphy, A. (1792) *An Essay on the Life and Genius of Samuel Johnson.* London.

Norden, J. (1618) *The Surveyor's Dialogue.* London.

Parker, A. J. (2001) 'Maritime landscapes', *Landscapes* 2, 122–41.

Parker, R. A. C. (1975) *Coke of Norfolk: A Financial and Agricultural Study, 1707–1842.* Oxford University Press, Oxford.

Parry, J. (2003) *Heathland*, National Trust, London.

Payne-Gallwey, R. (1886) *The Book of Duck Decoys: Their Construction, Management, and History.* Van Voorst, London.

Pedley, M. S. (1996) 'Maps, war and commerce: business correspondence with the London map firm of Thomas Jefferys and William Faden', *Imago Mundi* 48, 161–73.

Pedley, M. S. (2005) *The Commerce of Cartography: Making and Marketing Maps in Eighteenth-Century France and England.* University of Chicago Press, Chicago.

Pevsner, N. and Wilson, B. (2002) *The Buildings of England, Norfolk I: Norwich and North-East.* Yale University Press, London.

Phillips, A. D. M. (ed.) (1983) 'Introduction', in *A Map of the County of Stafford by William Yates, 1775.* Staffordshire Records Society 4th series, vol. 12, Stafford.

Plumb, J. H. (1952/3) 'Sir Robert Walpole and Norfolk husbandry', *Economic History Review* 5, 86–9.

Postgate, M. R. (1973) 'Field systems of East Anglia', in R. A. Baker and A. R. H. Butlin (eds), *Studies of British Field Systems.* Cambridge University Press, Cambridge, 281–324.

Prest, W. (1998) *Albion Ascendant: English History 1660–1815.* Oxford University Press, Oxford.

Rackham, O. (1986) *The History of the Countryside.* Dent, London.

Rackham, O. (1986) 'The ancient woods of Norfolk', *Transactions of the Norfolk and Norwich Naturalists' Society* 27:3, 161–77.

Ravenhill, W. L. D. (1965) 'Introduction', in *Benjamin Donn's Map of the County of Devon, 1765.* Devon and Cornwall Record Society New Series 9, University of Exeter.

Ravenhill, W. L. D. (1965) 'Benjamin Donn 1729–98: map-maker and master of mechanics', *Transactions of the Devonshire Association* 97, 179–93.

Ravenhill, W. L. D. (1972) 'Joel Gascoyne, a pioneer of large-scale county mapping'. *Imago Mundi* 26, 60–70.

Ravenhill, W. L. D. (1974) 'Introduction', in *Two Hundred and Fifty Years of Map-making in the County of Surrey: A Collection of Reproductions of Printed Maps Published Between the Years 1579–1823.* Harry Margary, Lympne Castle.

Ravenhill, W. L. D. (1992) 'Introduction', in *Christopher Saxton's 16th Century Maps: The Counties of England and Wales.* Chatsworth Library, Shrewsbury.

Ravenhill, W. L. D. and Padel, O. J. (1991) *A Map of the County of Cornwall Newly Surveyed by Joel Gascoyne.* Devon and Cornwall Record Society New Series 34, Plymouth.

Ravenhill, M. and Rowe, M. (2002) *Devon Maps and Map-Makers: Manuscripts and Maps before 1840.* Devon and Cornwall Record Society New Series 43, Exeter.

Rees, E. and Walters, G. (1970) 'The library of Thomas Pennant', *The Library, Transactions of the Bibliographical Society* 5th Series 25:2, 136–49.

Reeves, A. and Williamson, T. (2000) 'Marshes', in J. Thirsk (ed.), *Rural England: An Illustrated History of the Landscape.* Oxford University Press, Oxford, 150–66

Reynolds, A. (1997) 'The definition and ideology of Anglo-Saxon execution cemeteries', in D. Boe and F. Verhaeghe (eds), *Death and Burial in Europe*. Zelik, Bruges, 33–41.

Riches, N. (1937) *The Agricultural Revolution in Norfolk*. University of North Carolina Press, Chapel Hill.

Richeson, A. W. (1966) *English Land Measuring to 1800: Instruments and Practice*. Society for the History of Technology, Massachusetts Institute of Technology.

Robinson, A. H. W. (1972) *Marine Cartography in Britain*. Leicester University Press, Leicester.

Rodger, E. M. (1972) *The Large Scale County Maps of the British Isles 1596–1850: A Union List*, 2nd edn. Bodleian Library, Oxford.

Rodwell, J. S. (1991) *British Plant Communities Volume 2: Mires and Heaths*. Cambridge University Press, Cambridge.

Rogerson, A. (1995) *Fransham: An Archaeological and Historical Study of a Parish on the Norfolk Boulder Clay*. Unpublished PhD Thesis, University of East Anglia.

Roy, W. (1785) 'An account of the measurement of a base on Hounslow Heath', *Philosophical Transactions* 85, 2.

Saunders, H. E. (1937) 'Estate management at Raynham, 1661–1706'. *Norfolk Archaeology* 19, 39–76.

Sawyer, P. (1979) 'Medieval English settlement: new interpretations', in P. Sawyer (ed.), *English Medieval Settlement*. Edward Arnold, London, 1–8

Seymour, W. A. (ed.) (1980) *A History of the Ordnance Survey*. William Dawson, Folkestone.

Shaw, J. M. (1982) *Little Cressingham Mill*. Norfolk Windmills Trust, Norwich.

Sheail, J. (1971) *Rabbits and their History*. David and Charles, Newton Abbot.

Sheail, J. and Bailey, M. (1996) 'The history of the rabbit in Breckland', in P. Ratcliffe and J. Claridge (eds), *Thetford Forest Park: The Ecology of a Pine Forest*. Edinburgh, 16–20.

Silvester, R. (1988) *The Fenland Project, No. 3: Norfolk Survey, Marshland and the Nar Valley*. East Anglian Archaeology 45.

Skelton, R. A. (1952) *Decorative Printed Maps of the 15th to 18th Centuries*, Staples Press, London.

Skelton, R. A. (1962) 'The origins of the Ordnance Survey of Great Britain', *Geographic Journal* 128, 415–30.

Skelton, R. A. (1970) *County Atlases of the British Isles 1579–1850: A Bibliography 1579–1703*. Carta Press, London.

Skelton, R. A. (1974) *Saxton's Survey of England and Wales*. Nico Israel, Amsterdam.

Skempton, A. W. (1971–72) 'Early members of the Smeatonian Society of Civil Engineers'. *Transactions of the Newcomen Society* 44, 23–47.

Slack, P. (1990) *The English Poor Law, 1531–1782*. Macmillan, London.

Smith, A. H. (1956) *English Place Name Elements*, 2 vols. Cambridge University Press, Cambridge.

Smith, D. (1991) 'The Wild family firm', *Map Collector* 54, 32–8.

Smith, W. (1941) *The Land of Britain: Part 45, Lancashire*. Geographical Publications, London.

Sobel, D. (1996) *Longitude*. Fourth Estate, London.

Southwell, T. (1879) 'Norfolk decoys', *Transactions of the Norfolk and Norwich Naturalists' Society* 2, 538–55.

Spencer, J. and Thomas, R. (1992) *Norfolk Inventory of Ancient Woodland*. English Nature, Peterborough.

Stoker, D. (1979) 'An eighteenth century map piracy', *Norfolk Archaeology* 37:1, 123–6.

Strawhorn, J. (1959) 'Introduction', in *A Map of Ayrshire 1775 by A. and M.J. Armstrong*. Collections of the Ayrshire Archaeology and Natural History Society.

Stroud, D. (1965) *Capability Brown*. Country Life, London.

Sussams, K. (1996) *The Breckland Archaeological Survey*. Norfolk Museums Service, Suffolk County Council and English heritage, Ipswich.

Tarlow, S. (2007) *The Archaeology of Improvement in Britain, 1750–1850*. Cambridge University Press, Cambridge.

Tate, W. E. and.Turner, M. (1978) *A Domesday of English Enclosure Acts and Awards*. Reading University Library, Reading.

Taylor, C. (1973) *The Cambridgeshire Landscape*. Hodder and Stoughton, London.

Taylor, C. (1999) 'Post-medieval drainage of marsh and fen', in H. Cook and T. Williamson (eds), *Water Management in the English Landscape*. Edinburgh University Press, Edinburgh, 141–56.

Taylor, C. (2000) 'Fens', in J. Thirsk (ed.), *Rural England: An Illustrated History of the Landscape*. Oxford University Press, Oxford, 167–87.

Tooley, R. V. (1999) *Dictionary of Mapmakers*, revised edition by Josephine French. Map Collector Publications, London.

Turner, M. (1980) *English Parliamentary Enclosure: Its Historical Geography and Economic History*. W. Dawson, Folkestone.

Turner, M. (2005) 'Parliamentary enclosure', in T. Ashwin and A. Davison (eds), *An Historical Atlas of Norfolk*. Phillimore, Chichester, 131–2.

Turner, R. (1985) *Capability Brown and the Eighteenth-Century English Landscape*. Weidenfeld and Nicolson, London.

Tyacke, S. (1973) 'Map-sellers and the London map trade 1650–1710', in H. Wallis and S. Tyacke (eds), *My Head is a Map: Essays and Memoirs in Honour of R. V. Tooley*. Carta Press, London, 63–80.

Verner, C. (1959) 'Mr. Jefferson makes a map', *Imago Mundi* 14, 96–107.

'The Viatores' (1964) *Roman Roads in the South East Midlands, I.* Gollancz, London.

Vince, J. (1987) *Discovering Watermills*. Shire, Aylesbury.

Wade-Martins, P. (1990) *Village Sites in the Launditch Hundred*. East Anglian Archaeology 10.

Wade Martins, S. (1980) *A Great Estate at Work: The Holkham Estate and its Inhabitants in the Nineteenth Century*. Cambridge University Press, Cambridge.

Wade Martins, S. (1991) *Historic Farm Buildings*. Batsford, London.

Wade Martins, S. and Williamson, T. (1999) *Roots of Change: Farming and the Landscape in East Anglia 1700–1870*. British Agricultural History Society, Exeter.

Wailes, R. (1949–51) 'The windmills of Cambridgeshire', *Transactions of the Newcomen Society* 17, 97–119.

Walker, R. B. J. (2004) 'John Russell', in *The Dictionary of National Biography*. Oxford University Press, Oxford.

Walter, R. (1881) *An Index to Norfolk Topography*. Index Society, London.

Walters, G. (1968) 'Themes in the large scale mapping of Wales in the eighteenth century', *Cartographic Journal* 5, 135–46.

Watts, W. (1779) *The Seats of the Nobility and Gentry in a Collection of the Most Interesting and Picturesque Views*. London; repr. Garland, New York, 1982.

Welding, J. D. (1984) 'Introduction', in *Leicestershire in 1777: An Edition of John Prior's Map of Leicestershire*. Leicestershire Libraries and Information Services.

Whyte, N. (2005) *Perceptions of the Norfolk Landscape c.1500–1750*. Unpublished PhD thesis, University of East Anglia.

Williamson, T. (1993) *The Origins of Norfolk*. Manchester University Press, Manchester.

Williamson, T. (1995) *Polite Landscapes: Gardens and Society in Eighteenth-Century England*. Alan Sutton, Stroud.

Williamson, T. (1997) *The Norfolk Broads: A Landscape History*. Manchester University Press, Manchester.

Williamson, T. (1998) *The Archaeology of the Landscape Park: Garden Design in Norfolk, England, 1680–1840*. British Archaeological Reports, Oxford.

Williamson, T. (2000) 'Understanding enclosure', *Landscapes* 1:1, 56–79.

Williamson, T. (2003) *Shaping Medieval Landscapes*. Windgather, Oxford.

Williamson, T. (2007) *Rabbits, Warrens and Archaeology*. Tempus, Stroud.

Williamson, T. (2007) 'The landscape', in P. Cattermole (ed.), *Wymondham Abbey: A History of the Monastery and Parish Church*. Wymondham Abbey Book Committee, Wymondham, 172–85.

Withers, C. (2007) *Placing the Enlightenment: Thinking Geographically about the Age of Reason*. University of Chicago Press, Chicago.

Witney, K. (1990) 'The woodland economy of Kent, 1066–1348', *Agricultural History Review* 38, 20–39.

Wood, H. T. (1912) 'The Royal Society of Arts: the premiums. History of the premiums given for large scale county maps 1759–1809', *Journal of the Royal Society of Arts* 60, 263–74.

Woodward, D. (ed.) (1975) *Five Centuries of Map Printing*. University of Chicago Press, Chicago.

Worms, L. (1993) 'William Faden', in *The Dictionary of National Biography: Missing Persons*. Oxford University Press, Oxford.

Worms, L. (2000) 'Location in the London map trade', *International Map Collectors' Society Journal* 28, 32–42.

Worms, L. (2004) 'John Seller', in *The Dictionary of National Biography*. Oxford University Press, Oxford.

Worms, L. (2004) 'The maturing of British commercial cartography: William Faden (1749–1836) and the map trade', *Cartographic Journal* 41:1, 5–11.

Worms, L. (2004) 'Thomas Jefferys (1719–1771): beginning the world afresh', *Mapforum* 3 (Autumn), 20–9.

Yardy, A. (2004) *The Development of the Broadland Drainage Windmills, with Particular Reference to the Firm of Englands of Ludham*. Unpublished MA dissertation, School of History, University of East Anglia.

Yardy, A. and Scott, M. (2005) 'Windmills', in T. Ashwin and A. Davison (eds), *An Historical Atlas of Norfolk*, 2nd edn. Phillimore, Chichester, 172–3.

Yaxley, D. (ed.) (1988) *A Survey of the Houghton Estate by Joseph Hill, 1800*. Norfolk Records Society 30.

Yelling, J. A. (1977) *Common Field and Enclosure in England 1450–1850*. Macmillan, London.

Young, A. (1797) *General View of the Agriculture of Suffolk*, London.

Young, A. (1804) *General View of the Agriculture of Norfolk*. London.

Index

Entries in bold refer to the illustrations

Acle 147
Acts of Parliament 37
    Copyright Act 36
    Eau Brink Act 65
    Engraving Copyright Act 36
    General Drainage Act 158
    Longitude Act 22
Adams, George 23, 27, 38, 62
agricultural improvement 100, 132, 143
    reclamation 111
agricultural revolution 109
agriculture 123, 134, 143, 147, 156, 162
    arable cultivation 27–8, 30, 100,
        110, 111, 131, 134, 162, 173, 191
    fold course system 144, 148
    pasture 100, 162
    sheep–corn husbandry 144
Ainslie, John 22, 23, 25, 26, 50, 51, 60
alehouses see inns
American War of Independence 52,
    54, 91
André, Peter 18
Andrews, John 22, 24, 35
Ant, river 161, 170, 171
archaeological remains
    medieval 183
    prehistoric 183, 184, 185
    roman 183
Armstrong, Andrew 9, 19, 23, 32, 37,
    66–7
Armstrong, Mostyn 35, 66–9, 168
    his survey of Norfolk 9, 38, 49,
        67, 68, 83, 101, 139
Ashwellthorpe 190
Attleborough 183, 184
Aylsham Navigation, the 170, **171**,
    177
Ayrshire, maps of 31, 32
Badley Moor 113, 187

Baker, Henry 15
Banks, Joseph 92
barns 149, 151, **152**, 153
Barney 126
battlefields 29
Bawdeswell 180
Bayfield 138, 184
Becchamwell 184
Bedfordshire, maps of 50, 60
Beeston 129, 138, 184
Beetley 101
Begbie, Patrick 22, 34
Beighton, Henry 8, 21, 23, 25, 37,
    38, 39
Berkshire, maps of 18, 30, 52
Berwickshire, maps of 67
Besthorpe 115, 117
Billinge, Thomas 34
Bintry 145, 177
Bird, John 27
Blakeney 172
Blicking Park 129, 138, 179
Blomefield, Francis 127
    History of Norfolk 68
Board of Agriculture, the 18
Board of Ordnance 11–13, 18, 21, 35,
    41, 59, 62, 69, 91, 93, 94 *also see*
    Ordnance Survey
Bodney 111
Boughton 145
Brancaster 101, 102, 172, 183
Breckland 105, 109, 110, 124, 137,
    144, 148
Brettenham 145
brick kilns 174, **175**
brickmaking 174
bridges 172
Bridgeham 145, 148
Brisley Green 195

Briston 180
Broadland 122, 134, 153, 156, 157–8,
    159, 161, 172, 180
Bromholme 184
Broomhill 101
Brown, Lancelot 'Capability' 129, 138
Bryant, Andrew 11, 40, 88, 93
    his map of Norfolk 28, 30, 40,
        86, 88, 107
Buckenham 172
Buckenham Tofts 129
Buckinghamshire, maps of 50, 60
building materials 173–4
Burdett, Peter 9, 16, 19, 20, 34, 35,
    36, 38
Bure, river 153, 161, 170
Burgh Castle 183
Burnham Overy 172, 178

Camden's *Britannia* 6
Cary, John 22, 74, 83
castles 183, 184
Castle Acre 183, 184
Castle Rising 183
Catton 135
Cheshire, maps of 9, 19, 23, 24, 30,
    33, 34, 36, 40, 83, 90
Chet, river 170
chronometers 24
    churches 102, 115, 184–5
    as depicted on maps 87
    isolated 114–15, **116**
    ruined 184
Cley-next-the-Sea 178
coastal surveys 60, 64, **65**, 173
coastal trade 172
Coke, Thomas William 73, 132, 139,
    143, 151
Colby, Thomas 94, 96

Colkirk Common 195
commons 28, 100–103, 105, 106, 107, 108, 113, 114, 115, 125, 126, 128, 179, 185–7, 189, 191–2, 194–6, 200 *also see* common land
common-edge settlement 90, 102, 113–15
common fields 100, 143–9, **146**
distribution of 145, 147, 148
survival of 145, 147, 148, 149
common land 100, 157 *also see* commons; greens; heaths; intercommoning; moors; warrens; wetlands
amount of 103
and churches 115
as depicted on maps 87, 90
distribution of 102, **104**
and enclosed land 186
encroachments on 115, 117
and hundreds 186, 187, 189, **190**, 191, 196
'improvement' of 90
morphology of 103, 105, **106**
names of 107
and parish boundaries 103, **105**, 189–91, **190**, 191, 192
size of 105, 113
and soils 103, **104**, 107–8
terms for 106
and woods 195, 196
contour lines 31
copyright 36–7
Corbridge, James 8–9
Cornwall, maps of 6, 30
country houses
owners of 140–41
without park 140
county maps 1, 4, 6, 9, 17–18, 29–30, 200
accuracy of 83, 89
a 'cartographic revolution' 1, 17, 97
commercial viability of 4
complaints about 29
dates of **10**, **13**
dedications 38
financing of 9, 14, 37–8
numbers of 39
personal names on 38, 39, 139–40
plagiarism 6, 26, 74
price of 18, 40
production of 33–6

purchasers of 18–19, 29, 89, 143, 185
quality of 11, 26, 35, 36
reasons for production 14
scales of 30, 34
standardisation of 17
subscribers to 19, 38–9, 139, 140
time taken to complete 26
Coxford 184
Cranworth 192
Creake 184, 185
Crosley, William 13
Cumberland, maps of 9, 31, 39, 50, 52, 60, 140
Cunningham, William 4

Dalby, Isaac 93
Davis, Richard 19
Day, William 37
decoys 180–81
deer parks 129, 195
Defoe, Daniel 135
Derbyshire, maps of 9, 16, 20, 33, 38, 40, 74, 97
Dersingham 101
designed landscapes 30 *also see* landscape parks
Devon, maps of 19, 25, 26, 30, 36, 37, 39, 40, 50, 62, 74
Dickleburgh 117
Diss 127
Ditchingham 129, 180
Domesday Book 186, 191, 196
Donald, Thomas 9, 19, 23, 49, 50, 51, 60
Donn, Benjamin 16, 19–20, 25, 26, 30, 36
Dorset, maps of 16
Dunton Hall 140
Durham, maps of 9, 19, 32–3, 50, 67
Dury, Andrew 18, 22, 35, 60

Earsham 129
East Harling 145
East Wretham 145
economy, the 14, 165
Elmham Heath 195
Elmham Park 195
Emes, William 131
enclosure 100, 145, 148, 154 *also see* parliamentary enclosure
piecemeal 100, 144, 145

engravers 6, 14, 18, 21, 26, 29, 34, 36, 49, 51, 52, 85, 94
engraving 19, 21, 24, 29, 31, 34, 35, 36, 41, 57, 93, 97
Enlightenment, the 14, 18, 29
Essex, maps of 18, 39, 94
Evans, John 17

Faden, William
business activities 52, 54, 56–8
early life and career 50–52
engagement to survey Norfolk 68
family 50, 58
Geographer to the King 18, 52, 56
involvement in American mapping 34, 52–7, **55**
partnership with Thomas Jefferys junior 51
premises 51, **53**, 57, 59, **59**, 60
as publisher 13, 16, 21, 23, 49, 52, 54, 56–9, 93
purchase of copperplates by 38, 52, 58
retirement and death 60
takeover of Thomas Jefferys' business 21, 50, 52
Faden's map of Norfolk 49
appearance of 29, 97
boundaries on **150**
churches on 85–6, 89, 184
common fields on 145, **146**
common land on 28, 88, 89, 90, 100–102, 143, 185
cost of production 37
country houses on 86, 139
in digital form **2–3**, 78–9, 200
editions of 73
errors 85–6, 183–4
facsimiles of 78
financing of 39
geodetic accuracy of 80–81, 83, **84**, 97
geo-rectification of 80, 81, 83
landscape parks on 129, 131–3
'marsh and fen' on 102, 154, 158, 162
as military survey 69, **70**, 71
mills on 90, 161, 174, 177
orientation of 81, **82**, 83, **84**
price of 40
publication of 72–3
rivers on 86, 89

roads on 31, 89, 148, **150**, 165, **166**, 167, 184
scale of 34, 72, 73
settlement on 88, 89, 90
significance of date of 28, 101, 196, 200
time taken to survey 85
topographical accuracy of 86–9, 97, 131–2, 147
woodland on 86, 88, 89, 119, 122
farms 149, 151, 156
Felbrigg 129
Felthorpe 101
fen 156–8 *also see* common land
Fens, the 153, 154, 157, 159, 162
    draining of 158–9, 162–3
    northern ('Marshland') 113, 154–6, **155**, 159, 162
    southern 157, 159, **160**
ferries 172
Fersfield 127
field boundaries 148, 149
Fife and Kinross, map of 52
Flegg 101, 147
flooding 65, 156, 158
Foot, Thomas 94
Fordham, Herbert George 97
Forncett 145, 190
Foulsham 87
Foxley 89, 145
France, mapping of 15, 40–41, 92
free warren, grant of 110
Fritton Common 127, **127**, 128
Fulmondeston Common 195
Fulmondeston Severals 195
Fundenhall 190

Garboldisham 127
Gardner, William 13, 16, 20, 93
Gascoyne, Joel 6, 21, 30
Gasthorpe 145
Gawdy Hall 132
gibbets 187, **188**
Gloucestershire, maps of 83
Godwick 184
    Hall 140
Good Sands region, the 109
Gough, Richard 6, 16, 19, 67, 97
grain prices 100, 111, 123, 147, 148, 159, 162
Gream, Thomas 13, 16
Great Bircham 101, 102

Great Ellingham 114
Great Massingham 115
Great Melton 129
Great Yarmouth 170, 172, 179
    maps of 69
greens 113, 114, 115 *also see* common land
    as depicted on maps 87
    and soils 113
Greenwich Observatory 20, 24, 81
Greenwich–Paris Triangulation 27, 92
Greenwood brothers, the 9, 11, 40, 93
Gressenhall 117, 126, 180
Grimston 101
Guist 179
Gunthorpe 126
    Park 195

hachures 31, 79 *also see* relief
Halvergate Marshes 153, 154, 156, **157**, 161
Hampshire, maps of 16, 24, 52, 61
Hanworth 129
harbours 64, 172
Hardwick 115
Harpley 192
Harrison, John 24
Heacham 101
heaths 109–10, 114, 125, 126, 148 *also see* common land
    as depicted on maps 87
    enclosure of 122, 151
    ownership of 106
    and soils 109
    uses of 109
Heckingham 180
Helhoughton 192
Hemsby 180
Hertfordshire, maps of 6, 18, 30, 52
Hethel 129
Hethersett Common 191
Heydon 129
Hickling 184
'high farming' 122, 153
Hindolveston Common 195
Hingham 101
Hockering 195
Hodskinson, Joseph 9, 16, 21, 22, 35, 49, 50, 51, 60, 65, 74, 81, 83
Holkham 172, 185

Great or 'New' Barn 151, **153**
    Hall 139
    Park 124, 129, 131, **131**, 138, 151
Holt 179, 187
Hooghe, Cornelis de 4
Horning 172
Horningtoft Wood 195
Horwood, Richard 58, 62, 65
Houghton Hall 140
Houghton Park 129, 133, 138
Houses of Industry 117, 180 *also see* workhouses
Huddart, Captain Joseph 64–6, 81
hundreds 180, 186–7, 189, 191, 192
    boundaries of 30, 186, 187, 189, 196
Hunstanton 129
Huntingdonshire, maps of 38, 40, 50
hydrography 64

Ickburgh 110
Icknield Way, the 184
industry 31, 32–3, 173, 177
inns 168–9
intercommoning 103
interfluves *see* watersheds
Intwood Park 132, **132**, **133**

Jefferson, Thomas 57–8
Jefferys, Thomas 8, 21, 23, 24, 25, 35, 49–50, 51, 60, 67, 74, 81
    bankruptcy of 37, 50
Johnson, Samuel 50, 51

Kenninghall 127
Kent, maps of 4, 13, 71
Kent, Nathaniel 103
Kent, William 123, 124
Kilverstone 145
Kimberley 129, 138
King's Lynn 65, 94, 170, 172
Kirby, John 8

Lancashire, maps of 9, 19, 23, 26, 32, 33, 34, 38, 39, 40, 83, 90
landed estates 109, 121, 124–5
landowners 100, 144, 147, 148
landscape parks 129, 140
    age of 137, 138
    avenues 138
    distribution of **130**, 134
    lakes 133, 137, 138

landscape parks *continued*
   lodges 133
   proximity to towns 135, **136**, 137
   size of 129, **130**, 135, 137, 138
   and soils 134–5, 137
   and woodland 138
Land Utilisation Survey 62, 90–91
Langley 129, 138, 184
Larling 145
latitude and longitude 4, 22, 24, 81,
   83, 92
Leicestershire, maps of 9, 20, 23, 24,
   28, 33, 39
Letheringsett 179
lighthouses 64, 172
lime kilns 174, **175**
Lincolnshire, maps of 50
Lindley, Joseph 13, 20, 25, 26, 37, 39, 40
Litcham 101
Little Cressingham 178
Little Ellingham 101
Little Massingham 192
London, maps of 58, **59**, 61–2, **61**, 65
Lothians, the, maps of 67
Loudon, John Claudius 185
Lyng 177

magnetic variation 83
manors 186, 191
maps *also see* county maps
   enclosure 85, 87, 88, 90
   estate 85
   related to military activity 13, 14,
      18, 69, 91, 93
   of towns 4, 21, 49
map making 18, 91, 97
   on the Continent 18, 27, 31
   French 18, 27, 40–41
   map publishers 21–2
map trade
   Continental 6, 18, 51, 56
   in London 14, 21, 22
market places 169
marling 173
marsh 154 *also see* common land
   drained marsh 154, 161
   salt marsh 154, 156
Marshall, William 156, 157, 159
Martyn, Thomas 30
Masters, Charles 16, 26
Mautby 180
Melton Constable 195

Park and Hall 129, 138, **194**, 195
meridians 6, 24, 57, 81
Merton 129
Methwold 110, 111
   Heath 187
Middlesex, maps of 30
Mildenhall 110
milestones 31, 167–8, **169**
mills
   and commons 178
   distribution of 161–2, **176**, 177–8
   drainage 159, 161, 174
   water 32, 170, 177
   wind 32, 174, 178–9
Milne, Thomas 16, 19, 23, 49, 51,
   60–63, 81
   London Land Utilisation Map
      61–2, **61**
minsters 191
moors 106, 113, 115 *also see* common
   land
   and soils 113
Morden, Robert 6, 24
Mousehold Heath 103, 105, 189
Mudge, Lt. William 93, 94
Mulbarton 115
'multiple estates' 191

Napoleonic Wars 100, 122, 147, 148
Nar, river 170
Narborough 183
New Buckenham 183
New Houghton 133
Norfolk, maps of **4**, **7**, 8–9, **8**, 11, 28,
   30, 40, 86, 88 *also see* Faden's map
   of Norfolk; Bryant, Andrew
Norfolk Broads *see* Broadland
Northamptonshire, maps of 50
North Elmham 140, 195
North Tuddenham 101
Northumberland, maps of 19, 37, 66
Norton Subcourse 145
Norwich 135, **136**, 137, 169
   maps of 4, 6

Ogilby, John 6, 34
Old Buckenham 101
open fields *see* common fields
Orbury Common 195
Ordnance Survey 4, 12, 24, 31, 35, 51,
   62, 81, 94, 96, 165 *also see* Board
   of Ordnance

maps by 11, 17, 27, 29, 30, 71, 80,
   86, 87, 88, 89, 94, **96**, 129
Ouse, river 65, 159, 163
   Great 158, 170
   Little 170, 177
Oxborough 172
Oxfordshire, maps of 19, 30, 50

parishes 189–90, 192, **193**
   boundaries of 30, 189, 191
parliamentary enclosure 27, 89, 143,
   145, 148, **151**
   chronology of 27–8, **28**, **29**, 100,
      149
   removal of common land by
      27–8, 100–101, 162
parochiae 191
Paterson, Daniel 74, 83
Peebleshire, maps of 35, 67
Pentney 184
pits 173, **175**
pollards 126, 127, **128**, 174
poor houses 179, 180, 186
poor laws 117, 179, 180
Post Master General, the 18, 74
prehistoric earthworks 29
Priestley, Rev. Joseph 65
Prior, John 9, 19, 20

Quartermaster General, the 69, 74, 94

Ramsden, Jesse 26–7, 41, 51, 92
Ranworth 145
Raynham 129, 138
Reedham 172
relief 31, **32**, 91 *also see* hachures
religious houses 184
Rennie, John 65
Repps 147
Reymerston 192
Reymerston Common 192
Reynolds, Captain Thomas 69
Richmond, Nathaniel 129, 138
Riddlesworth 145
Ringstead 101, 173
roads 30–31, 74, 155, 165 *also see*
   turnpike roads; Roman roads
   as depicted on maps 87, 88, 89, 148
   improvements to 165
   maintenance of 165, 167
   major 165, 167–9
   minor 165

private 165
Roman 184
unbounded 148
road books 6, 34, 67, 83
Rocque, John 18, 21, 22, 24, 29, 30, 35
Rollesby 180
Roy, General William 27, 62, 91–2
Royal Society, the 15, 21, 64–5, 91–2
Rudham, East and West 192
Runham 145, 147
Rushford 145
Ryburgh Wood 195

Saxon period, the 102, 183, 186, 187, 191, 196
Saxton, Christopher 4, 13, 14, 21, 31
Sayer, Robert 60, 64
Scarning 101
Scotland, maps of 19
Scratby 147
Seller, John 6, 24
settlement 154–6
    common-edge *see* common-edge settlement
    as depicted on maps 87, 88, 89
    deserted or shrunken 184
    dispersed 149
    drift 102, 114 *also see* common-edge settlement
Seven Years War, the 41, 49, 52
Shernbourne 173
Shipdham 126, 128, 192
Shottesham 184
Shouldham 184
Shropshire, maps of 24
Smallburgh 180
Smeaton, John 50, 51, 92
Smeatonian Society, the 26, 51, 61, 64, 93
Society of Arts, the 15, 16–17, 20, 26, 29, 37, 50, 58, 67, 73–4, 89
    prizes awarded by for county maps **10**, 15–16, 18, 20, 25, 33, 38, 61, 66, 97
soils 107, 145
    chalky 108, 109, 124, 134, 148
    heavy clays 108, 113, 114, 115, 123, 126, 134–5, 191, 192
    light clays 108, 113, 115, 135, 192
    loams 108, 109, 113, 115, 124, 134–5, 192
    peat 115, 134, 157, 159, 162

sands 108, 109, 111, 113, 115, 123, 124, 126, 134–5, 137, 148, 192
silts 113, 115, 134, 162
soil maps 62
Somerset, maps of 26, 29, 34, 37, 40
South Burgh 192
Southery 172
South Walsham 145
Speed, John 4, 6
Staffordshire, maps of 9, 19, 23, 24, 33, 36, 38, 72, 89
Stanford 111
state involvement in cartography 13–14, 41
Stiffkey 153
Stiphens Green 195
Stock Heath 103, 126, 189, **189**, **194**, 195
Stokesby 101
Stow Bardolph 129
Strachey, John 29, 37
Strumpshaw 172
Sturston 111
Suffolk, maps of 8, 9, 16, 38, 39, 49, 52, 81, 83
Surlingham 172
Surrey, maps of 8, 18, 20, 25, 26, 30, 37, 39, 40, 74
survey
    geodetic 22–5, 81, 83, 93
    topographical 25–6, 93, 94
    trigonometrical 6, 9, 94
surveying instruments 9, 14, 23, 25, 26–7
    Great Theodolite, the 27, **63**, 92
surveying methods 9, 96
    textbooks on 23
    triangulation 9, **11**, 12, **12**, 23, 81, 91
surveyors 19–21, 22, 64
    amateur 16, 17–18, 19–20
    estate 8, 16, 20, 29
Sussex, maps of 8, 13, 16, 30, 37
Swaffham 170
Swanton Common 195
Swanton Morley 101, 177
Swanton Novers 88, 126
Great Wood 195
Swinnow Wood 195

Tacolneston 145, 190
tan offices 179
Tasburgh 185
Taverham 177, 187

Taylor, Isaac 16, 20, 24, 29, 35
Taylor, Samuel 185
Telford, Thomas 18
territories 191–2
Testerton Wood 195
Thelveton 101
Thetford 110, 111, 177, 183
Thorpe Parva 101
Thurgarton 179
Thurne, river 156, 161, 170
Thursford 126
Thury, Cassini de 15, 21, 91
Tiffey, river 153, 187
Titchwell 101
Tottington 101, 111
towns 135, 139, 141, 169, 170, 179
transport 15, 165 *also see* roads
    improvements in 14, 135
    by water 170–72
Trigonometrical Survey, the 12, 59, 69, 92, 93, **95**
Trinity House 64
tree-planting 124
turnpike roads 30, 135, 165, 167, **168**
    gates 31, 167
    tolls 31, 167
    turnpike trusts 14, 85, 165

Upwell 145

vegetation types 109, 113, 119
Venta Icenorum 183
Vermuyden, Cornelius 158, 159

Wales, maps of 8, 17, 39
Walpole, Horace 133
Walpole St Andrew 101
Walsingham 184
Warham 185
warrens 106, 110–12, **112**, 148 *also see* common land
    lodges 110–11, **111**
    ownership of 106, 110
    and soils 113
Warwickshire, maps of 8, 9, 21, 23, 33, 37, 38, 72
watercourses 158, 170–72
    locks 170, 171, 177
    navigable 170
    new 65–6, **66**, 158, 159, 163, 170, 172
    staiths 172

watersheds 103, 105, 137, 178, 187, 189, 191, 196
water supply 114
Watton 117
Waveney, river 170
Wayland Wood 186
Weasenham St Peter 192
Wells-next-the-Sea 172, 179
Welney 172
Wensum, river 153, 177
West Bradenham 187
West Harling 145
Westmorland 39, 50, 60
West Raynham 192
Westwick Park 137
wetlands 153 *also see* fen; Fens, the; marsh *also see* common land
Whinburgh 192
Whittlingham 172
Whyman, Joseph 20, 23
Wickhampton 145
Wicklewood 117, 180
Wighton 178
    Hall 140

Wilton 172
Wiltshire, maps of 24, 31, 39
Windham, Joseph 73
Winfarthing Oak 185
Winterton 147, 180
Wisbech 94
Wissey, river 170
Wolterton 138
Woodbastwick 145
woodland 89, 138
    amount of 123
    ancient 119, 121, 123, 194, 195–6
    carr 122
    and commons 195
    coppice-with-standards 119, 121, 195
    as depicted on maps 89
    destruction of 122
    distribution of **120**, 123, 196, **197**, **198**
    management of 119, 123, 195
    plantations 121–2, 123
    recent 121, 123–5

    shapes of 123
    and soils **120**, 123, 124, 196
    uses of 121
Wood Norton 180
wood pastures 125–6, 195–6, **197**
    on commons **125**, 126–9, 192, 194, **197**
Woods, Richard 129
workhouses 179, 180
    as depicted on maps 87
Wretham 111
Wyatt, Samuel 151
Wyld, James 59
Wymondham 117, 191–2

Yare, river 153, 161, 170
Yates, William 9, 19, 23, 24, 26, 34, 35, 36, 38
Yeakell, William 13, 16, 20
Yorkshire, maps of 30, 31, 32, 33, 38, 50, 52, 60, 64
Young, Arthur 162